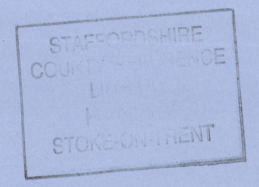

Ceramic-Matrix Composites

Ceramic-Matrix Composites

Edited by

R. WARREN

Department of Engineering Metals
Chalmers University of Technology
Göteborg
Sweden

Blackie
Glasgow and London

Published in the USA by
Chapman and Hall
New York

Blackie and Son Ltd.
Bishopbriggs, Glasgow G64 2NZ
and
7 Leicester Place, London WC2H 7BP

Published in the USA by
Chapman and Hall
a division of Routledge, Chapman and Hall, Inc.
29 West 35th Street, New York, NY 10001-2291

© 1992 Blackie and Son Ltd
First published 1992

British Library Cataloguing in Publication Data

Ceramic matrix composites.
I. Warren, R.
620.1

ISBN 0-216-92682-3

Library of Congress Cataloging-in-Publication Data

Warren, R. (Richard)
  Ceramic-matrix composites / R. Warren.
    p.   cm.
  Bibliography: p.
  Includes index.
  ISBN 0-412-02021-1
  1. Fibrous composites. 2. Ceramic fibers. I. Title.
TA418.9.C6W37 1990
620.1'4—dc20
                                                            89-17270
                                                            CIP

Typeset by Keytec Typesetting Ltd, Bridport, Dorset
Printed in Great Britain by St Edmundsbury Press, Bury St Edmunds, Suffolk

Preface

The past decade has seen a rapid development in ceramic materials towards their use in structural applications. Improved processing in combination with better microstructural design (based on improved understanding of mechanical behaviour) has led to materials with sufficient strength and reliability being developed for commercial applications. In particular, ceramics are expected to extend the upper limits of high temperature components.

In view of the progress made, as well as the challenges facing future development, it is appropriate that the state of current knowledge be reviewed. The objective of this book is to review a particularly promising form of ceramics, namely ceramic composites. Here, composites are defined as all ceramics consisting of two or more ceramic microstructural constituents combined to improve mechanical properties. Fibre-containing materials are of obvious interest but equal attention is paid to particulate composites which include the important transformation toughened ceramics. Composites with metallic constituents (e.g. cermets and cemented carbides) are not considered. Functional ceramics (e.g. insulation, superconducting materials, coatings, etc.) are only discussed where a secondary load-bearing capacity is demanded.

Following the introduction, four chapters provide a foundation for an understanding of ceramic composites. Chapter 2 reviews the many fibres and whiskers that are available as potential reinforcements in ceramic matrices, whilst chapters 3 and 4 present the fundamental principles of composite preparation and properties. These chapters provide a summary of current knowledge and theoretical understanding, and where theoretical models are described they are kept as simple as possible (suitable references are given for more rigorous treatments). In a similar manner chapter 5 introduces the reader to the principles to be considered when designing with ceramic composites.

Chapters 6–9 deal with important classes of composites, namely particulate composites, long-fibre reinforced ceramics, chemical vapour infiltrated fibre composites and whisker reinforced ceramics. These chapters discuss details of processing and properties of specific materials in order to provide a basis of assessment for materials engineers as well as extending the principles treated in the foundation chapters.

The authors contributing to this book all have in-depth experience in the research and development of ceramic composites, having devoted

time both to their exploitation in structural applications and to fundamental aspects. It is our hope that this book will be of value as an
introduction to the subject, both to practising materials scientists and
advanced students.

RW

Contributors

Dr. A. R. Bunsell   Ecole Nationale Supérieure des Mines de Paris, Centre des Matériaux Pierre-Marie Fourt, B.P. 87, 91003 Evry Cedex, France

Dr G. C. Eckold   Engineering Division, Harwell Laboratory B424, United Kingdom Atomic Energy Authority, Oxfordshire OX11 0RA, UK

Professor J. Homeny   Department of Ceramic Engineering, University of Illinois, Urbana IL 61801, USA

Dr R. Lundberg   Volvo Flygmotor AB, S-46181, Trollhätten, Sweden

Professor R. Naslain   University of Bordeaux, Laboratory for Thermostructural Composites (UMR 47 CNRS-SEP-UB1), Europarc, 1 Avenue Léonard de Vinci, 33600 Pessac, France

Dr D. C. Phillips   Research Laboratory, Kobe Steel Europe Ltd, 10 Nugent Road, Surrey Research Park, Guildford, Surrey GU2 5AF, UK

Dr V. K. Sarin   Division of Engineering, Boston University, Massachusetts, 02215, USA

Dr R. Warren   Department of Engineering Metals, Chalmers University of Technology, S-41296 Göteborg, Sweden

Contents

5 Structural design with advanced ceramic composites   115

G. C. ECKOLD

1 Overview

R. WARREN

1.1 Introduction

Until recently, materials for advanced structural applications have been largely restricted to advanced metallic alloys. It now appears, however, that for certain applications, above all those involving high temperatures, metallic materials have reached a limit in their potential for development. In combustion engines and other energy generating equipment, the desirability of higher operating temperatures has driven the development of Ni-, Co- and Fe-base alloys to extreme levels of sophistication in microstructural and structural design. However, such development is necessarily limited by the melting point of these metals. For continued development, ceramics offer one of the few avenues to a significant increase in service temperatures. Their low density, chemical inertness and high hardness offer additional potential for extending performance limits beyond those offered by metallic materials. Hitherto, the widespread use of ceramics has been inhibited by their brittleness and poor reliability of strength. In an effort to overcome these problems, considerable progress has been made during the past two decades in the compositional and microstructural design of ceramics. Among the most important advances has been the application of the composite principle, i.e. the combination of two or more constituent phases with appropriate microstructural morphology in a material, thereby achieving an improved combination of properties.

This chapter provides a brief preview of the subject of structural ceramic composites, their applications and potential as well as the main scientific and technical principles underlying their development and properties.

1.2 Scope and definitions

Ceramics can be defined as chemically very stable, inorganic, crystalline, non-metallic compounds or mixtures of such compounds. Their chemical bonding is generally a hybrid of covalent and ionic, the proportions of which vary between compounds. The characteristic properties that are the result of such bonding are for example high melting point, high chemical stability, high elastic moduli and low dislocation and atomic mobility, the latter leading to high hardness and creep resistance but also to brittleness. Ceramics are compounds consisting of at least one element from early in the periodic system and they consequently usually have relatively low density.

The bounds set by the above definition are not strict. A large number of compounds with high melting points, notably the carbides, nitrides, borides and silicides of the transition metals are borderline ceramics. They often exhibit metallic character being electrically conducting and having poor chemical resistance. They are nevertheless of interest as constituents in ceramic composites. The element carbon, because of its extremely high temperature potential is also of interest in spite of its poor oxidation resistance. Important examples of ceramics and related compounds used in currently developed composites are listed with some of their significant properties in Tables 1.1 and 1.2.

Ceramic composites can be considered to be materials consisting of two or more distinct ceramic phases combined on a microstructural scale. In the context of the present work the combination is chosen and produced intentionally to give an optimised set of properties not achievable in monolithic ceramics. In general, the properties of a composite are not simply an average of those of the constituents but are sensitive to details of the microstructure as will be outlined below and as is made clear in the other chapters of the book. It is this potential for

Table 1.1 Selected physical properties of monolithic, polycrystalline ceramics and refractory compounds

Material	Density (g/cm^3)	Melting point (°C)	E (GPa)	v	C_p (J/g K)	Thermal conductivity (W/m K) (1000 °C)	α (K^{-1}) (0–1000°)	Electrical resistivity (Ω m; 25 °C)
Al_2O_3	3.99	2050	390	0.23	1.25	6.0	8.0	$> 10^{15}$
SiC	3.2	$\approx 2500^a$	440	0.15	1.25	40	4.5	≈ 1
Si_3N_4	3.2	1900^a	300	0.22	1.25	15	3	–
B_4C	2.5	2450	440	0.18	2.1	15	5.5	0.5
BN (cubic)	3.5	$\approx 3000^a$	–	–	2	–	–	–
BN (hex.)⊥	2.3	–	45	–	–	21	7.5	10^{11}
BN (hex.)∥	2.3	–	70	–	–	14	0.8	–
AlN	3.26	2300^a	320	0.25	1.1	50	6	2×10^{12}
TiB_2	4.5	2980	570	0.11	1.23	25	5.5	10^{-5}
TiC	4.9	3070	450	0.18	0.85	30	8.5	10^{-4}
TiN	5.4	3090	–	–	0.85	≈ 30	8.5	5×10^{-5}
$MoSi_2$	6.25	2100	400	0.17	0.56	20	8.5	2×10^{-5}
ZrO_2 (tetr.)	6.1	2400^a	–	–	0.7	2	12	–
ZrO_2 (mono)	5.55	–	240	0.3	–	–	15	–
Mullite	2.8	1850	150	0.24	1	5	5.5	–

[a]Dissociates or transforms.

Table 1.2 Selected mechanical properties of monolithic polycrystalline ceramics and refractory compounds

Material	K_{Ic} (MN m$^{-3/2}$)	Hardness (VPH; GPa)		Wear resistance[a] $K_{Ic}^{3/4} H^{1/2}$	Thermal shock resistance[a] $[(1 - v)K \times K_{Ic}/E\alpha] \times 10^3$
		20 °C	1000 °C		
Al_2O_3	3–5	19	7	10–14.5	5.9
SiC	3–4	26	–	11.5–14.5	60
Si_3N_4	3.5–6	16	12	10–15	65
B_4C	3.7	45	20	17.9	20
BN (cubic)	–	≈ 50	–	–	–
BN (hex.)	–	≈ 5	–	–	–
AlN	2.7	13	–	7.6	53
TiB_2	5–7	30	–	21	43
TiC	4	30	4	15.5	30
TiN	–	20	–	–	–
$MoSi_2$	5	12.5	–	11.8	24
ZrO_2 (tetr.)	5–10	12	–	11.5–19.5	4
ZrO_2 (mono)	1	–	–	–	–
Mullite	2	15	10	6.5	9

[a]Figures of merit (see chapter 4).

microstructural engineering that has made ceramic composites one of the main topics of materials science research and development in recent decades.

It is necessary here to distinguish between so-called functional ceramics and structural ceramics. The former have been developed and are used primarily for non-mechanical properties such as thermal insulation, electrical resistance/conductivity, etc. whereas the latter have been developed for applications requiring capacity to bear various forms of mechanical loading, for example high-temperature strength, and wear and erosion resistance. This book is concerned primarily with structural ceramic composites; properties other than mechanical properties are considered but only where these are relevant to the function of materials developed for applications demanding significant load-bearing capacity.

1.3 The constituents of ceramic composites

Examples of the most important constituent compounds of current ceramic composites are listed in Table 1.1 together with some relevant properties. Risking overgeneralisation, it can be stated that the most important monolithic ceramics developed as structural materials are aluminium oxide, zirconium oxide, silicon carbide and silicon nitride. The properties of these have been further improved either by alloying (e.g. silicon nitride with alumina to form sialons or zirconia with other oxides such as yttria, ceria or magnesia to give transformation micro-structures) or by using the composite principle. In the latter case the four ceramics are used as matrices to which second phase constituents are added in the form of particles, fibres or platelets. Various types of glass and glass-ceramics are of interest as the matrices of composites.

As has already been implied, the primary purpose of the above-mentioned modifications to structural ceramics is to improve the mechanical properties, above all the toughness (see section 1.4) but also for example hardness and creep strength. Nevertheless, it is important to note that other properties such as thermal conductivity and thermal expansion can have vital significance for the mechanical performance under certain conditions (e.g. thermal shock). Secondary functional properties such as the electrical and thermal properties can also be of decisive significance in specific structural applications.

The way in which the mechanical properties of composites are influenced by the properties and morphology of their constituents are presented briefly in the next section and in greater detail in later chapters of the book.

1.4 Strength, toughness and reliability

The ceramic materials of interest here rarely exhibit plastic deformation below at least 1000 °C. Their strength is consequently determined by the catastrophic extension of a crack developed from an internal flaw. This can be expressed by the fracture mechanics equation:

$$\sigma_F = Y K_c/c^{1/2} \tag{1.1}$$

where Y is a dimensionless constant dependent on the geometry (not size) of the flaw and the geometry of the stress field and the sample, c is the flaw size and K_c is the so-called fracture toughness. An equivalent but perhaps more revealing expression applicable to extension of a crack under tensile load in plane strain is

$$\sigma_F = Y[\Gamma E/c(1 - v^2)]^{1/2} \tag{1.2}$$

where Γ is the fracture surface energy, E the Young's modulus and v the Poisson ratio. This reveals that the strength decreases not only with a decreasing resistance to crack formation of the material (i.e. loosely speaking its bond strength) and with increasing flaw size but also with decreasing elastic stiffness.

The practical implications of equations (1.1) and (1.2) are that strength can be improved either by reducing flaw size in the material or by improving the toughness. Thus, significant progress has been made in improving the processing of ceramics in order to reduce the frequency and size of defects (see chapter 3). The benefit of this approach is however limited since in defect-free material intrinsic features of the microstructure act as flaws, e.g. grains or grain boundaries that crack during loading. Moreover, brittle materials are sensitive to surface damage; consequently, a material with a low level of defects may be subsequently weakened by defects accumulated during service.

The second approach, i.e. the improvement of toughness, can be achieved most effectively at the microstructural level using a variety of mechanisms such as crack deflection, crack induced phase transformation, controlled microcracking and fibre reinforcement. Such mechanisms are discussed in detail in later chapters. A preliminary illustration is given in Figure 1.1 which shows two distinct types of toughening obtainable in composites. The first is deflection of various kinds in which the growing crack is led around second phase particles or fibres. This causes a reduction in stress intensity at the crack tip and therefore an apparent toughness increase. The second type is observed in ceramics reinforced with high volume fractions of long, parallel fibres; here, cracks pass through the matrix leaving intact fibres bridging them. If the fibres have a higher stiffness than the matrix then the matrix will crack not only at a proportionally higher composite stress than the unreinforced matrix but also at a somewhat higher strain, i.e. it will be toughened. After failure of the matrix, loading of the composite can continue until fibre failure. This normally occurs by successive fracture and pull-out of individual fibres leading to pseudo-plastic behaviour giving relatively high fracture energies.

Another important implication of equations (1.1) and (1.2) is that, since defects are usually distributed stochastically with respect to size and location, there will be a corresponding scatter in strength from specimen to specimen. Moreover, the mean strength of a series of specimens or components will be volume-dependent, i.e. it will decrease with increasing component size. These effects are illustrated in Figure 1.2 which covers ranges of strengths typical for currently available ceramics.

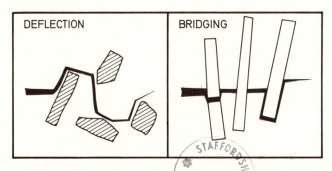

Figure 1.1 Schematic illustration of deflection and fibre bridging mechanisms.

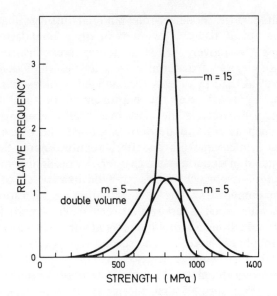

Figure 1.2 Strength distributions typical for ceramic materials with different degrees of scatter (m is the so-called Weibull modulus, an inverse measure of the width of the distribution, see chapters 4 and 5).

To a designer, such strength scatter translates as poor reliability and necessitates the use of very conservative strength criteria. Although the effect can be mitigated by reducing defect populations through improved processing, sensitivity to subsequent damage must be considered. In certain circumstances strength scatter is reduced by increasing toughness (see chapter 4). The long fibre type of toughening described above is particularly valuable in this respect.

1.5   Important microstructural parameters

Against the background of the above discussion of strength and toughness, a number of microstructural parameters that are relevant to toughening can be identified:

- volume fraction of constituents
- shape of second phase, e.g. particulate, platelet or fibre
- dimensions of second phase constituents, i.e diameter, length and aspect ratio
- orientation of fibres with respect to the loading direction

In addition to these purely geometrical factors it will be shown that the relative magnitudes of the *thermal expansion, elastic and mechanical properties of the constituents* as well as the *properties of the interface* between them are critical in determining the nature of fracture. For example, the success of mechanisms of fibre toughening such as crack deflection and crack bridging is dependent on the existence of a relatively weak fibre/matrix interface as well as a correct balance of elastic properties. The relative values of the expansion coefficients of the constituent phases determines the level and sign of residual stresses in the microstructure. Although it can be understood that this affects such processes as crack deflection and microcracking, the overall effect on strength and toughening is not easily predicted. It is probable that a second phase with a somewhat higher thermal expansion than the matrix has a beneficial effect since this results in an average compressive stress in the matrix after cooling following processing. Values of expansion coefficients are included in Table 1.1.

1.6 Examples of ceramic composites

Examples of microstructural geometries commonly found in ceramic composites are shown in Figure 1.3 while specific examples of composites that have been produced and reported on are listed in Table 1.3. The fibres mentioned in the table are described in further detail in chapter 2. It can be noted that the number of reinforcement/matrix combinations is relatively limited. This is partly because development has been largely restricted to matrices of the most well-known ceramics, partly because of the limited number of available fibres and partly because many combinations are chemically incompatible.

Early attempts to reinforce ceramics included the use of refractory metal wires [1]. An expected benefit was the augmentation of the fracture energy by the plastic deformation of the fibres. A problem was

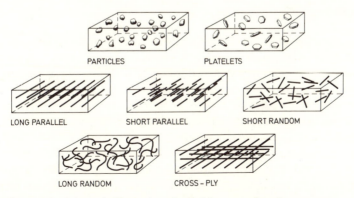

PARTICLES   PLATELETS

LONG PARALLEL   SHORT PARALLEL   SHORT RANDOM

LONG RANDOM   CROSS-PLY

Figure 1.3   Schematic illustration of principle composite microstructures.

Table 1.3   Examples of tried ceramic/ceramic composites

	Composite type and constituents matrix–reinforcement	Details in chapter
Particulate	Al_2O_3–ZrO_2 Al_2O_3–TiC Al_2O_3–SiC	6
	SiC–TiB_2 Si_3N_4–TiC Si_3N_4–ZrO_2	6
Platelets	Al_2O_3–SiC (pl) Si_3N_4–SiC (pl)	6
Short random fibres (Whiskers)	Al_2O_3–SiC (w) Si_3N_4–SiC (w)	9
Long, parallel fibres	glass–C glass–SiC	7
Cross-plied	glass–C glass–SiC SiC–SiC	7 8
Woven	C–C SiC–SiC	8

found to be the recrystallisation and consequent weakening and embrittlement of the wires during processing. Refractory metals are also very sensitive to oxidation. That a ductile metal can contribute to the toughness of brittle solids is however demonstrated by the long-established class of materials known as cemented carbides (e.g. WC-Co alloys) and the idea has been resurrected recently in the form of ceramics combined with metallic particles. Such ceramic/metal composites are not treated in this book.

1.7   Compatibility

The preparation of ceramic composites generally necessitates consolidation at very high temperatures. Consequently, potential constituents cannot be combined with retained integrity if they react chemically, are intersoluble or form low melting point eutectics. Even if a preparation process can be devised to minimise the interaction, the problem may arise during subsequent service at elevated temperatures. An exhaustive list of such interactive systems cannot be given here. The compatibility of a given combination of materials can in most cases be predicted by reference to existing experimentally determined phase diagrams or by thermodynamic estimates based on free energies of formation and reaction [2]. Relevant examples of interactions are:

(1) $2Si_3N_4 + 2ZrO_2 = Zr_2ON_2 + 3Si_2N_2O$ (chemical reaction)
(2) $Si_3N_4 + 3C = 3SiC + 2N_2$ (chemical reaction)
(3) $ZrO_2 + xY_2O_3$ = solid solution; x up to ≈ 1 (dissolution)
(4) $ZrO_2 + Al_2O_3$ = eutectic melt $\geqslant 1800\,°C$ (eutectic formation)

In certain circumstances such interactions, if appropriately controlled can be exploited to create composite microstructures *in situ* (see chapter 3).

1.8 Processing and manufacture

The general principles of preparation and manufacture of ceramic composites are presented in chapter 3 while details for specific composite types are found in the later chapters. Briefly it can be stated that particulate composites can be prepared using powder processing technology developed for traditional ceramics, i.e. mixing of the constituents in powder form followed by a forming stage and then sintering. Forming and sintering can be carried out simultaneously if a hot pressing process is used although this generally involves greater expense. Although whisker reinforced ceramics can be prepared with similar techniques, pressing, forming and sintering are made more difficult by the presence of whiskers; in general, a hot pressing consolidation process is necessary. This and the higher cost of whiskers compared to powder makes the whisker reinforced ceramics 20–40% more expensive than particulate composites [3].

For composites reinforced with long, parallel fibres, special techniques have had to be developed. One method involves impregnating continuous multifilament yarns or tapes with a matrix-powder slurry. The slurry contains a binder which holds the impregnated tape together after drying. The tapes can then be stacked to form the desired component dimensions and cross-ply geometry. Consolidation normally requires hot pressing. An alternative technique involves impregnation of a fibre preform with a matrix by means of a vapour-to-solid chemical reaction that uses the fibre surfaces as a substrate for the deposition of the solid reaction product. The most common examples are SiC/SiC and C/C composites. The multidirectional fibre preforms are usually prepared from multifilament yarn by weaving techniques [4]. To achieve sufficiently low porosity the impregnation has to be repeated several times with intermediate surface machining; the process is therefore very expensive. An alternative is to impregate the preforms with a solution or liquid organic precursor which deposit the matrix during a subsequent pyrolysis.

A number of novel, alternative techniques have been developed recently involving *in situ* chemical reactions. These are presented in chapter 3.

1.9 Applications and future potential

As implied above, particulate and whisker reinforced ceramics have superior properties to monolithic ceramics and are therefore replacing these in many applications where the somewhat higher production costs can be offset by improvement in performance. In such applications as cutting and forming tools, wear parts in machinery, nozzles, valve seals and bearings, improvements in toughness and hardness translate into longer life and a reduced frequency of delays caused by catastrophic failure. Where required, composites can be chosen to exhibit thermal and electrical conductivity, e.g. to permit spark machining. Perhaps the most successful single application of whisker and particulate composites is as cutting tool inserts (Figure 1.4) (see also chapter 6). The current market for cutting tools in the United States is estimated at around 30 million dollars, the total for all applications of ceramic composites being around 100 million dollars.

Figure 1.4 Ceramic and ceramic composite cutting tool inserts (by courtesy of Greenleaf Corporation, Saegertown PA, USA).

An important driving force for the development of ceramic composites has been the potential for their use in combustion engines and other energy conversion systems. Above all, an increase in the operating temperatures of such equipment would lead to increased efficiency and reduced exhaust emissions. Although the feasibility of producing and running such components as combustion chambers, catalyser substrates, engine valves, impeller and turbine wheels has been demonstrated (see chapter 8), poor reliability and high cost are still inhibiting factors.

A second area of potential application still awaiting improved reliability is biological implants; ceramic materials generally show excellent compatibility with biological tissue and fluids.

Long-fibre composites offer a higher mechanical reliability but have so far not achieved the necessary microstructural stability at the high temperatures necessary in combustion engines and other high temperature applications. The main reasons for this appear to be (a) structural degradation or oxidation of fibres at high service temperatures and (b) incompatibility of the constituent phases in oxidising environments at high service temperatures. The most significant current example of an application of this type of composite is the use of carbon/carbon composites as brake materials in aircraft brake systems.

Currently, considerable research activity is being devoted to developing fibres with higher stability (e.g. continuous, monocrystalline fibres) and to finding compatible ceramic combinations. At present, it seems likely that these will be found among the refractory oxides. Another approach that is being pursued is to find a protective coating system for C/C composites to prevent oxidation.

References

1. Buljan, S. T. and Sarin, V. K. (1987) Silicon nitride-based composites, *Composites* 18, 99–106.
2. Singh, M. and Wiedemeier, H. (1989) Thermomechanical modelling of interfacial reactions in intermetallic and ceramic matrix composites, in: *Interfacial Phenomena in Composite Materials*, Butterworths, London, pp 303–309.
3. Karpman, M. and Clark, J. (1987) Economics of whisker-reinforced ceramics, *Composites* 18, 121–124.
4. Ko, F. (1989) Preform fiber architecture for ceramic matrix composites, *Am. Ceram. Soc. Bull.* 68, 40–414.

2 Ceramic fibres for reinforcement

A. R. BUNSELL

2.1 Introduction

Ceramic fibres may be divided into three very different types, namely, (i) those made by chemical vapour deposition (CVD) onto a substrate, (ii) those made by the conversion of a precursor fibre into a ceramic fibre and (iii) whiskers. The first of these was developed in the early 1960s and is characterised by a large diameter, usually around 140 μm, good mechanical properties but inherent high cost. The boron and silicon carbide fibres made by the CVD process have been produced on both tungsten and pitch based carbon fibre cores [1]. This type of fibre is not generally considered for the reinforcement of ceramics as their large diameters lead to relatively few fibres bridging any incipient crack in a ceramic matrix and this is important in the toughening of this type of material [2].

The 1960s also saw the development of carbon fibres, the first inorganic fibre to be produced via a polymeric precursor fibre. In the early 1970s carbon fibres were successfully incorporated into glass matrix composites (see chapter 7). Small diameter ceramic fibres began to appear in the later 1970s with the development of the α-alumina fibre FP from Du Pont [3]. These fibres have diameters of around 20 μm. It was the development of fine diameter silicon carbide based fibres by the pyrolysis of polycarbosilane precursor fibres however which stimulated research into reinforcing ceramics. These fibres have diameters generally between 10 and 20 μm and result from research begun in the 1970s [4]. The Nicalon fibre which resulted from this research first became commercially available around 1982. There now exist several types of small diameter ceramic fibres which are candidates for reinforcing ceramic matrices and their number is likely to increase. Carbon fibres with diameters 10 μm or less made from the pyrolysis of polyacrylonitrile fibres or by the conversion of pitch precursor fibres are also possible candidates for reinforcing ceramics, especially pitch based fibres due to their greater resistance to oxidation [5]. The affinity of carbon for oxygen at high temperatures is however a problem which must be overcome, possibly by coating the fibre.

Ceramic whiskers which are filamentry monocrystals with diameters generally less than 1 μm and lengths from 0.1 mm to several millimeters are also potential reinforcements for ceramics. The strength of whiskers is extremely high, approaching the theoretical limit and their potential for reinforcement has been recognised for many years [6]. However the fear that they represent a health hazard, the difficulty of aligning the whiskers and so achieving optimum properties and their generally high cost has so far limited their exploitation.

This chapter will treat small diameter ceramic and carbon fibres as well as whiskers. Emphasis will be placed on the preparation and structure of the fibres and their properties in isolation; their behaviour in specific ceramic matrices will be described in other chapters.

One of the most important fibre properties is fracture strength. Since ceramic fibres are brittle their fracture is characterised by a large scatter in strength. In this chapter the strength values quoted are mean values based on representative samples of many individual fibres. The scatter of strengths can be described in statistical terms, the most common approach being based on Weibull statistics as described in chapter 4.

2.2 Fine alumina and mullite fibres

Alumina is usually produced by heating aluminium hydroxide or other aluminium salts, such as ammonia alum to above 1000 °C in air. During the heating process, the starting materials lose water and other fugitive components at 400–500 °C and are thus converted to aluminium oxide (Al_2O_3). This undergoes several transformations of crystal structure, and finally attains the most stable form of alumina namely α-alumina. All other transitional alumina are transformed into α-alumina at 1000–1100 °C. In the α-alumina structure, oxygen atoms are positioned in a hexagonal close-packed structure and the interstices are filled regularly and densely with aluminium atoms. α-Alumina normally occurs as relatively large crystalline grains, generally larger than 0.1 μm. It has a high degree of crystallinity, high density (3.97 g/cm^3), a stable and inactive surface, and is strong and hard.

When polycrystalline α-alumina is heated above 1400 °C, the grains grow larger and sintered structures with grains having diameters up to 100 μm are formed at 2000 °C. It melts at 2300 °C.

The first small diameter ceramic fibre which became available in 1979 from Du Pont de Nemours consisted of almost pure α-alumina and was called Fiber FP [7]. This fibre is made by blending the alumina in powder form with water and other compounds to form a mixture which can be spun. The filaments are then extruded through spinnerets, drawn and fired in two steps. It is believed that an initial heating in air to 1300 °C, with careful control of shrinkage, produces an α-alumina fibre and this is followed by heating in a propane air flame for a few seconds which coats the fibre with a thin layer of silica [3, 8]. The silica layer increases fibre strength by healing surface flaws and also helps wettability with molten metals. Fiber FP is granular with grain sizes of about 0.5 μm. Properties of the FP fibre are shown in Table 2.1. It can be seen that the fibre has a low strain to failure which makes it difficult to handle and this limits its use. However the FP fibre retains its strength well up to 1000 °C showing almost no loss after 300 h when heated to this temperature in air [3]. Above these temperatures however grain growth occurs making the fibre even more brittle and generally weakening it.

As a candidate for reinforcing glass and ceramic-matrix composites, Du Pont has developed a fibre known as PRD-166. This fibre is described as being microcrystalline based on α-alumina but containing approximately 20% by weight of partially stabilised zirconia [9]. The role of the zirconia is said to be to stabilise the structure so improving strength and toughness. As can be seen from Table 2.1 this results in a considerable improvement over the properties of the FP fibre. Figure 2.1 shows results published by Du Pont indicating that the PRD 166 fibre also has greatly enhanced tensile strength at high temperature compared to fiber FP. It should be noted however that these results were obtained at room temperature after heat treatment for 2 h. Heating the PRD-166 fibre in air to 1400 °C produces a 35% fall in

Table 2.1 Properties of alumina and mullite based fibres

Manufacturer	Fibre type	Typical composition (wt%)	Tensile strength (GPa)	Strain to failure (%)	Young's modulus (GPa)	Specific gravity	Diameter (μm)
Du Pont de Nemours	Fiber FP	>99 α-Al$_2$O$_3$	1.40	0.4	380	3.9	20
Du Pont de Nemours	PRD-166	80 α-Al$_2$O$_3$ 20 ZrO$_2$	2.07	0.6	380	4.2	20
Sumitomo Chemicals	Alf	85 Al$_2$O$_3$ 15 SiO$_2$	2	1.1	180	3.2	18
ICI	Safimax	96 δ-Al$_2$O$_3$ 4 SiO$_2$	2.0	0.7	300	3.3	3
3M	Nextel 312	62 Al$_2$O$_3$ 24 SiO$_2$ 14 B$_2$O$_3$	1.75	1.1	154	2.7	11
	Nextel 440	70 Al$_2$O$_3$ 28 SiO$_2$ 2 B$_2$O$_3$	2.1	1.1	189	3.05	11
	Nextel 480	70 Al$_2$O$_3$ 28 SiO$_2$ 2 B$_2$O$_3$	2.3	1.0	224	3.05	11

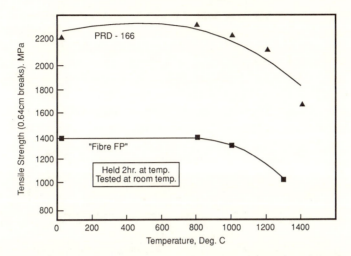

Figure 2.1 Comparison of the strengths as a function of temperature of Fibre FP and PRD-166, both fibres by E. I. du Pont de Nemours (after Romine [9]).

strength almost immediately but this appears to stabilise so that no further fall in strength is observed after 100 h [9]. The enhanced high temperature strength seems to be developed through zirconia transformation toughening. This is said to be achieved by preventing the thermodynamically favoured transformation of zirconia from the tetragonal to the monoclinic phase during cooling after the sintering process. The tetragonal phase is said to be stabilised by the addition of yttria or other rare earth oxides. The PRD-166 fibre is polycrystalline, the α-alumina crystals having a size of approximately 0.5 μm and the zirconia grains about 0.1 μm.

The properties of two other alumina fibres are shown in Table 2.1. Both the Sumitomo fibre and the ICI Safimax fibre contain silica but in differing amounts. Figure 2.2 shows the effect of silica on the Young's modulus of alumina fibres and Figure 2.3 shows its effect on strength [10]. It can be seen that increasing the silica content of alumina fibres reduces the fibre modulus whereas the strength increases.

The γ-alumina fibre from Sumitomo Chemicals (Alf) contains 15% by weight of silica and, as Table 2.1 shows this leads to a much lower Young's modulus than that of the α-alumina fibres but a greater strain to failure and therefore easier handleability.

The fibre is made by starting with an organo-aluminium compound such as trialkyl aluminium or trialkoxy aluminium which is then polymerised by the addition of water to give polyaluminixane [11]. This organo-metallic polymer is then dissolved in an organic solvent containing alkylsilicate to give a viscous product which can be spun to a fibre.

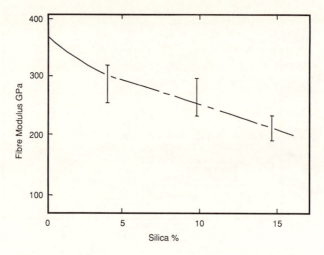

Figure 2.2 The elastic modulus of alumina fibres falls as they are intimately combined with increasing percentages of silica [10].

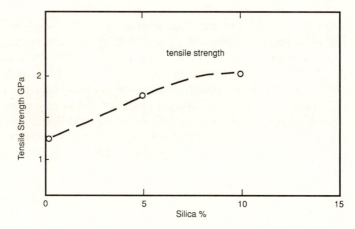

Figure 2.3 The strength of alumina fibres increases as they are intimately combined with increasing percentages of silica [10].

The resulting precursor fibre is then heated in air to 760 °C which carbonises the organic groups to give a fine ceramic fibre composed of alumina and amorphous silica. The fibre is then heated to 970 °C which causes the alumina to crystallise in the γ form with a very fine grain size, a low density and an active surface. The role of the silica, which makes up 15 wt% of the structure, is described by the maker as being to stabilise the γ form of alumina. In common with most other small diameter ceramic fibres the Sumitomo alumina fibre shows considerable spread in diameters around the average value of 18 μm. This is most probably due to the difficulty in spinning the precursor fibres.

Figures 2.4 and 2.5 reveal that the fibre conserves its strength and stiffness at least up to 800 °C with little noticeable loss in tensile properties at 1000 °C. Above this temperature the tensile properties fall and at temperatures above 1150 °C the fibre deforms plastically [12].

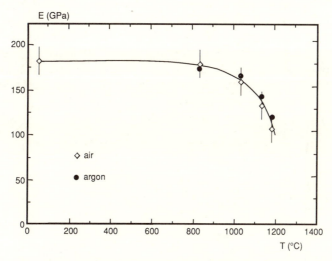

Figure 2.4 The Young's modulus of the alumina silica fibre from Sumitomo Chemicals is maintained constant until at least 800 °C.

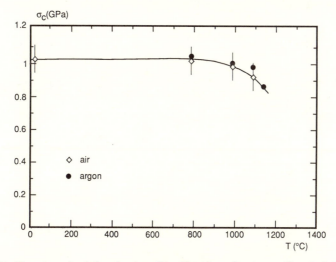

Figure 2.5 The strength of the alumina silica fibre from Sumitomo Chemicals begins to fall above 800 °C.

Above 1000 °C the fibre is observed to creep. Loading to a steady level produces an elastic strain followed by a period of primary creep and then secondary creep during which the creep rate is a linear function of time, as described by equation (2.1).

$$\frac{d\varepsilon}{dt} = K\sigma^n \exp(-\Delta H/RT) \tag{2.1}$$

where ε is the creep rate, σ is the applied stress and ΔH the activation energy of the creep mechanism. This behaviour is shown in Figure 2.6. The value of n depends on the type of creep mechanism predominating and values for this parameter as a function of temperature and atmosphere are shown in Table 2.2. A value for n of around 2 has been shown to be due to a diffusion controlled creep process occurring by an interfacial reaction at grain boundaries [13].

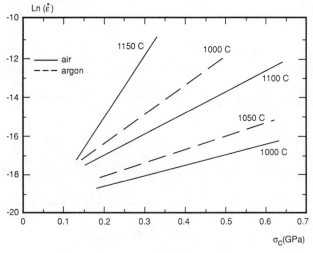

Figure 2.6 Creep rate of the alumina silica fibre from Sumitomo Chemicals as a function of applied stress at different temperatures.

Table 2.2 Variation of the exponent of the stress for the creep of the Sumitomo alumina fibre

Temperature	Atmosphere	n
1050	Argon	3.3
1100	Argon	4.4
1000	Air	2.2
1100	Air	2.1
1150	Air	16.0

X-Ray and electron diffraction analysis of the as-received fibre confirm that the alumina is crystalline γ and that the silica is amorphous. Subsequent heat treatments produce changes in the diffraction pattern with the appearance of mullite at and above 1127 °C. The rate at which the structure is converted to mullite is strongly temperature dependent. At 1127 °C, 5–10% of the structure is converted after 100 h whereas the conversion to mullite was completed after 2 min at 1400 °C. A residue of α-alumina exists after the conversion to mullite since the silica content is insufficient for complete reaction.

The as-received fibre consists of γ-alumina crystals with dimensions around 10 nm although some as large as 25 nm are observed. A dark field electron micrograph of the as-received structure is shown in Figure 2.7 and should be compared with the micrograph in Figure 2.8 of a fibre after conversion to mullite revealing large crystals of up to 0.3 μm.

The fibre is found to be remarkably inert to interaction with the surrounding atmosphere; its elemental composition does not change during heating in air or in argon. However the fibre is limited in use at high temperature by the development of mullite. The change of structure which occurs at a relatively low temperature and the fall in modulus beginning at 800 °C is probably caused by the softening of the amorphous silica phase surrounding the microcrystalline alumina particles.

The Safimax fibre is a long staple fibre with a diameter of 4 μm produced by ICI and consisting of about 96% alumina and 4% silica.

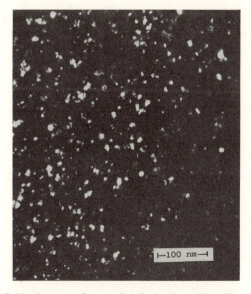

Figure 2.7 Dark field electron micrograph of alumina silica fibre as received from Sumitomo Chemicals showing crystals of approximately 12 nm.

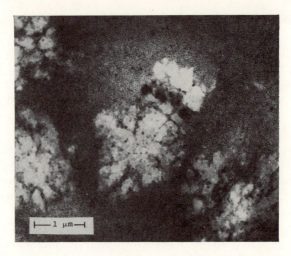

Figure 2.8 Dark field electron micrograph of the alumina silica fibre from Sumitomo Chemicals after 100 h at 1127 °C in air showing large mullite crystals surrounded by much smaller γ-alumina grains.

Such fibres are produced by solution spinning which leads to little scatter in diameter. The solution consists of an aluminium salt, such as aluminium chlorohydroxide $AlC_x(OH)_y$, which contains hydrated species which eliminate water to link octahedral units by Al–OH–Al and Al–O–Al bridges [14, 15]. As the process continues, a viscous solution is obtained which can be fiberised. This is helped by the addition of organic polymers to control rheology and to lower the spinning temperature. Progressive increases in temperature burn off the organic material and convert the alumina salt to alumina. The Saffil fibre which is a shorter form of the Safimax fibre was originally produced as a refractory felt and has been described as being fiberised by an extrusion-blowing technique in which the liquid is extruded through apertures into a high velocity gas stream. During conversion the fibre is initially highly porous and this porosity is removed by a heat treatment at up to 1600 °C. To avoid rapid conversion to α-alumina with consequent entrapment of the porosity as well as large crystals, both of which would lead to loss of mechanical strength, a small percentage of silica is deliberately introduced into the structure which inhibits the phase transformation. As α-alumina forms, the silica crystallises as mullite, the presence of which controls the growth of the alumina crystals at grain boundaries. The final structure consists of a mixture of δ- and α-alumina with a predominance of the former in the form of crystals of about 50 nm.

It is believed that prolonged heating of Saffil or Safimax fibres above 1000 °C leads to a gradual change from the delta to the alpha structure.

Complete conversion of the structure at 1200 °C is estimated to require years of heating [15]. At temperatures higher than 1200 °C shrinkage occurs but it is reported for early versions of the fibres that even at 1550 °C the shrinkage was less than 4% after 100 h. Saffil and Safimax fibres have been used mainly in metal-matrix composites. The fibres are shaped into preforms of desired shape and infiltrated with the molten matrix alloy, usually an Al alloy.

The Nextel series of fibres produced by 3-M are described as having a microcrystalline mullite structure. The three grades commercially available, Nextel, 312, 440 and 480, are said to possess a structure based on a 3 to 2 mole ratio of alumina to silica with additional boron oxide, 14% by weight for the Nextel 312 fibre and only 2% for the other two fibres. The crystallinity of the fibres increases in the order of their reference numbers. The manufacturer indicates that all three fibres conserve at least 75% of their tensile properties up to 1000 °C (Figures 2.9 and 2.10). It is to be noted that the strength variations are slightly different in different environments. Figure 2.11, again taken from reports of the manufacturer reveals that the Nextel fibres creep at 1000 °C and above.

The currently available alumina fibres seem to be possible candidates for reinforcing light alloys with the observation however that their specific gravities are quite high. Such reinforcements permit metal structures to be made with enhanced properties, notably creep resistance at high temperatures. The phase change to α-alumina above about 1000 °C and subsequent grain growth would however seem to exclude these fibres from use in most ceramic-matrix composite materials.

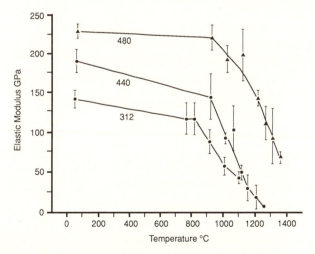

Figure 2.9 Variation of elastic moduli of the three Nextel fibres as a function of temperature (3-M Data).

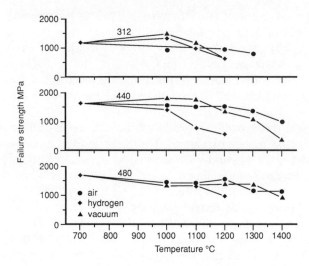

Figure 2.10 Variation of the strength of Nextel fibres as a function of temperature (3-M results).

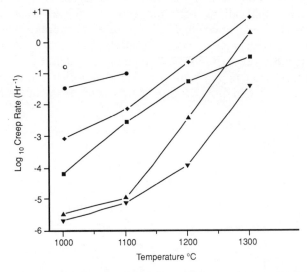

Figure 2.11 The creep rates of Nextel fibres (3-M results). ○ 312, 69 MPa; ● 312, 138 MPa; ■ 440, 69 MPa; ◆ 440, 138 MPa; ▼ 480, 69 MPa; ▲ 480, 138 MPa.

2.3 Silicon carbide and related fibres

The work of Yajima and his colleagues was first published in Japan in the mid-1970s and gave rise to the first and so far the most successful fine ceramic fibre based on silicon carbide [16, 17]. This fibre is now produced by Nippon Carbon under the name of Nicalon. Its manufacture involves the production of a polycarbosilane precursor fibre. Polydimethylsilane is made by reacting sodium with dichlorodimethylsilane.

$$SiCl_2(CH_3)_2 \xrightarrow{Na} (Si(CH_3)_2)_n$$

This is then heated in an autoclave at a pressure of about 10 MPa which results in a reorganisation of the polymer and the introduction into the chain of Si–C giving a polycarbosilane

$$(Si(CH_3)_2)_n \longrightarrow \left[\begin{array}{c} CH_3 \\ | \\ Si-CH_2 \\ | \\ H \end{array} \right]_n$$

The structure of the polycarbosilane consists of cycles of six atoms arranged in a similar manner to the cubic structure of β-SiC. The molecular weight of this polycarbosilane is low however, around 1500, which makes drawing of the fibre extremely difficult. In addition methyl groups (CH_3) in the polymer are not included in the Si–C–Si chain so that during pyrolysis the hydrogen is driven off leaving a residue of free carbon. The precursor fibres are subjected to heating in air at about 300 °C to produce cross-linking of the structure. This oxidation makes the fibre infusible but has the drawback of introducing oxygen into the structure which remains after pyrolysis. The ceramic fibre is obtained by a slow increase in temperature in an inert atmosphere up to 1300 °C. The fibre which is obtained contains a majority of SiC but also significant amounts of free carbon and excess silicon combined with oxygen. The distribution of these elements is uniform across the fibre diameter, as has been shown by electron microprobe analysis [18].

The strength and Young's moduli of Nicalon fibres tested in air or an inert atmosphere show little change up to 1000 °C. Above this temperature both these properties fall with the greatest change being in the strength of the fibre [18, 19] . Figure 2.12 shows the change in strength with temperatures of a number of different Nicalon fibres.

Nicalon fibres as currently supplied by the manufacturer are microcrystalline with typical grain sizes of the β-SiC crystals of around 2 nm, as is revealed by X-ray diffraction and transmission electron microscopy. Heating above 1100 °C causes grain growth which stabilises at an average grain size of approximately 3 nm. Under no load, the fibre shrinks during such a heat treatment but the shrinkage ceases with the stabilisation of the grain size [20].

When a load is applied to the fibres it is found that a creep threshold

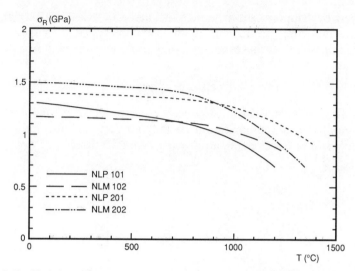

Figure 2.12 Variation of strength of several types of Nicalon fibres produced by Nippon Carbon.

stress exists above which creep occurs. The primary creep stage is clearly greatly influenced by the structural changes occurring. This can be seen from Figure 2.13 which shows how the primary creep behaviour of an early Nicalon fibre was influenced by heat treatment. Nicalon fibres produced later were more microcrystalline and showed less

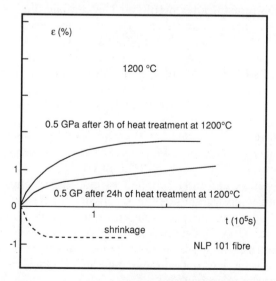

Figure 2.13 The early HLP 101 Nicalon fibres were found to shrink on heating under no applied load but to creep when under load. Heat treatment at 1200 °C reduced the primary creep.

primary creep at high temperatures. The ceramic grade 200 series Nicalon fibres which evolved from these earlier fibres has clearly been heat treated in order to stabilise grain size at around 3 nm and improve the properties at high temperature. The β-SiC crystals are embedded in an amorphous matrix of silica which also contains agglomerations of free carbon with sizes of around 2 nm. A typical composition by weight of a Nicalon fibre is 65% SiC, 15% free carbon and 20% SiO_2 as shown in Table 2.3. The free carbon and the oxygen in the structure have been shown to be limiting factors in the use of Nicalon fibres at high temperatures.

The 200 series ceramic grade fibres show creep stress threshold levels, as can be seen from Figure 2.14. The existence of the stress thresholds is attributed to the presence of the particles of carbon in the structure, which impede its movement and rearrangement. In this case the carbon particles confer a useful property to the fibre but their presence, together with the oxygen in the structure, finally limits the use of the fibre at high temperature. Figure 2.15 shows how the paramagnetic electron-spin resonance signal produced by the free carbon in the Nicalon fibre falls with heat treatment, indicating that it is being consumed probably by combining with the oxygen. It is significant that the rate of fall in intensity is slower in air than in an inert atmosphere. This is interpreted as being due to the creation of a silica coating on the Nicalon fibre when it is heated in an oxidising atmosphere. This oxide coating is considered to slow the expulsion of the oxides of carbon produced in the interior of the fibre.

Table 2.3 Properties of fine ceramic fibres based on silicon

Manufacturer	Fibre type	Typical composition (wt%)	Tensile strength (GPa)	Strain to failure (%)	Young's modulus (GPa)	Specific gravity	Diameter (μm)
Nippon Carbon	Nicalon	65 SiC 15 C 20 SiO_2	2.7	1.4	185	2.55	15
Ube Chemicals	Tyranno	Si, C, O Ti < 5	3	1.5	200	2.4	9
Dow Corning/ Celanese	MPDZ	47 Si 30 C 15 N 8 O	1.9	1.1	180	2.3	12
Dow Corning/ Celanese	HPZ	59 Si 10 C 28 N 3 O	2.2	1.5	150	2.35	10
Dow Corning/ Celanese	MPS	69 Si 30 C 1 O	1.2	1.6	190	2.65	11

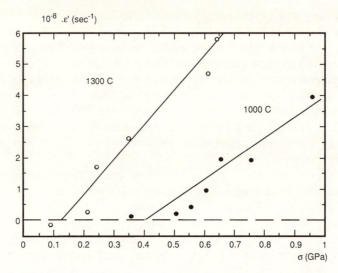

Figure 2.14 Nicalon fibres show distinct creep threshold levels above which the fibres creep. The threshold level decreases with increasing temperature.

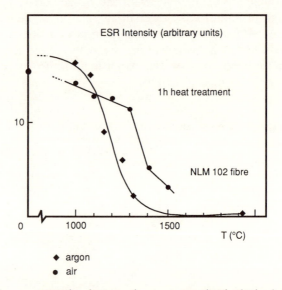

Figure 2.15 The paramagnetic electron-spin resonance signal obtained from the free carbon phase in the Nicalon fibre can be seen to decrease with heat treatment but at a different rate in argon (♦) than in air (●).

The influence of the environment on the behaviour of fibres can be considerable and should not be neglected. It has been observed that Nicalon fibres in a silicon carbide matrix resist temperatures above 1000 °C better than the isolated fibre. On the other hand such a ceramic-matrix composite can become embrittled, due to fibre/matrix interaction [20]. Nicalon fibres in a pure aluminium matrix have been seen to react with the molten matrix material at temperatures as low as 700 °C [21].

The Nicalon fibres are currently the most widely available fine ceramic fibre based on silicon and carbon but modifications of the technology of this process of fibre production are being actively explored in many laboratories [22].

The Tyranno fibre produced by Ube Chemicals is made by a process analogous to that used for the Nicalon fibre though with significant differences. Its manufacture begins with the production of polydimethysilane, $(Si(CH_3)_2)_n$, as for the Nicalon fibre; this is then mixed with titaniumalkoxide in a ratio of approximately 1:10, heated to 340 °C in nitrogen and polymerised for 10 h [23]. The reaction product is concentrated at 310 °C for 2 h in a stream of nitrogen to produce polytitanocarbosilane. The molecular weight of this organo-metallic polymer is low, around 1500.

The structure of the polytitanocarbosilane precursor is complex due to condensation of Si–H bonds and cross-linking of titanium compounds occurring simultaneously [23]. The resulting unit structure consists principally of

$$
\begin{array}{ccc}
\overset{|}{C}H_2\ (R) & \overset{|}{C}H_2 & \overset{|}{C}H_2 \\
CH_3-\overset{|}{\underset{|}{Si}}-\overset{|}{O}-Ti-O & CH_3-\overset{|}{\underset{|}{Si}}-H & CH_3-\overset{|}{\underset{|}{Si}}-CH_3 \\
(R) & &
\end{array}
$$

$(R = C_nH_{2n+1})$ with small amounts of other products.

The precursor fibre is converted into an amorphous ceramic fibre by heating in air at about 180 °C and then heat treating in nitrogen at around 1000 °C. The resulting fibre contains a small amount of titanium, around 2% by weight which is said to inhibit crystallisation, and similar percentages of SiO_2, C and SiC as the Nicalon fibre. The fibre has a diameter of around 8 μm.

Figure 2.16 shows the strength of Tyranno fibres as a function of temperature. Above 1200 °C the amorphous structure of the fibre changes and it becomes crystalline.

A related group of ceramic fibres which are being actively studied are based on a silicon carbide–silicon nitride system and are produced from polycarbosilazane precursors. Various routes for the manufacture of these fibres are possible and as yet they are at the laboratory or early pilot plant stage. The fibres can be produced by the pyrolysis of polysilazanes prepared from tris(N-methylamino) methylsilane polymerised by heating to 520 °C for several hours and then heated to between 1200 °C and 1500 °C for 2 h [24]. During pyrolysis a weight loss of approximately 35% is reported to occur. The $Si_x\ N_y\ C_z$ ceramic fibres

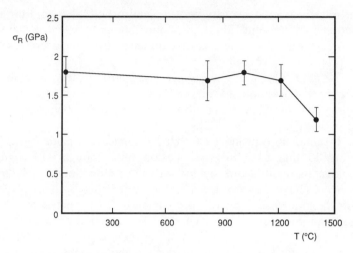

Figure 2.16 The strength of Tyranno fibres made by Ube Chemicals as a function of temperature.

which are finally produced are described as being shiny black in appearance [25].

This type of fibre was studied initially because of the dielectric properties which it is possible to obtain with this class of material. The electrical resistivity of the silicon carbide–silicon nitride combination is about 10^{12} times greater than that of graphite. Relatively high mechanical properties have been reported in the scientific literature for these fibres with moduli around 200 GPa and breaking strengths of about 1 GPa [26]. No data are available for their properties at high temperatures.

Other approaches to producing fibres based on Si–C–N–O have been studied and reported in the literature by researchers from Dow Corning and Celanese [26]. These fibres were prepared by melt spinning amorphous thermoplastic polymers followed by a cross-linking step to gel the fibres and to make them infusible. The Si–C–N–O fibres were derived from methylpolydisilylazane (MPDZ) prepared by reaction of methylchlorodisilane and hexamethyldisilazane. Fibres consisting of Si–N–C were derived from hydridopolysilazane (HPZ) polymers prepared by reaction of trichlorosilane and hexamethyldisilazane. The fibres were pyrolysed in nitrogen or argon at temperatures between 1100 °C and 1400 °C. The ceramic fibres produced were amorphous; their room temperature properties are given in Table 2.3 [28]. Also reported by the same research team is a fibre (termed MPS) described as being 50–80% microcrystalline β-SiC. This fibre, described as at an early stage of development, can be made with a stoichiometric SiC structure [27]. No high temperature data are available for these fibres but they are being developed specifically to reinforce ceramic matrices [28] and it is reported that they should be able to operate above 1000 °C [29].

2.4 Carbon fibres

Carbon fibres are able to withstand very high temperatures without loss of strength up to and above 3000 °C but only in the absence of oxygen. When heated in air, carbon fibres made from polyacrylonitrile precursors are observed to lose weight and oxidise. The lowest temperature usually given in the literature for this to occur is 400 °C. However weight loss has been observed as low as 250 °C [30]. The term carbon fibre covers a large family of fibres which may possess very different properties and as already mentioned can be made by one of two possible commercial routes [31]. Carbon fibres used for advanced composites have Young's moduli between 240 GPa and 800 GPa. Those most widely used at present have moduli at the bottom of this range, generally between 240 GPa and 400 GPa, and fracture strengths of 2–4 GPa.

The most common carbon fibres, i.e. those used with resin matrices are made from an organic precursor, polyacrylonitrile (PAN). This precursor is very similar to acrylic textile fibres but can be slightly modified with small percentages of co-polymers and above all with great attention paid to purity of the product. Figure 2.17 shows the various stages of the conversion of PAN to a carbon fibre. Important stages are the cross-linking of fibre in air at about 250 °C and the maintaining of an applied load on the fibre during heating to prevent shrinkage and pyrolysis in an inert atmosphere at temperatures above 1000 °C. The temperature of pyrolysis determines the final properties with maximum

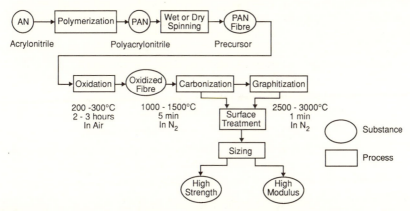

Figure 2.17 Schematic description of the conversion of polyacrylonitrile fibres to carbon fibres (after Toray Industries).

strength achieved at around 1500 °C and increasingly higher moduli obtained as the temperature is raised from 1500 °C to 3000 °C. The carbon yield obtained by this method is around 45%.

The structure of carbon fibres made from PAN is anisotropic with the carbon groups arranged more or less parallel to the fibre axis but with complete orientational disorder in the cross-section [32]. The structure of this type of fibre is never graphitic even in the high moduli fibres. The fibres contain considerable porosity which however is less in the higher moduli fibres. The surface reactivity of the fibres decreases with increasing modulus reflecting the better order at the surface of the high modulus fibres. Carbon fibres made from PAN were successfully used to reinforce glass as long ago as the early 1970s [33] and have more recently been used in carbon–carbon composites. However, as already mentioned they suffer from oxidation when heated in the presence of oxygen, which generally severely limits the service temperature of the composite (see chapter 7).

Carbon fibres produced by the conversion of pitch obtained either as a residue of oil refining or the coking process involves the purifying of the pitch and its conversion into a precursor fibre which is then heated in a process similar to that of the PAN based fibres. The pitch is first converted by one of several techniques into a mesophase or liquid crystal solution which contains phases in which the carbon groups are directionally ordered [31]. Passing the mesophase pitch through a spinneret produces a precursor fibre which has considerable molecular orientation parallel to the fibre axis.

Carbon fibres produced from mesophase pitch can possess extremely high Young's moduli with values of up to 800 GPa. This is because the structure of the fibres can be better organised than is the case with the PAN based fibres and a true graphite structure can be achieved. The fibres are consequently weaker in compression than their PAN based counteparts but they are also less reactive at the surface which improves their oxidation resistance and stability in matrices.

The prevention of oxidation of carbon fibres by some coating acting as an oxygen barrier seems to be a solution for producing a reinforcement which could be used at very high temperatures, possibly up to 2000 °C. Difficulties to be overcome are in ensuring a continuous unbroken protective layer over the whole of the fibre and the avoidance of exposure of the fibre for example due to machining.

2.5 Whiskers

Single ceramic crystals in the form of filaments are known as whiskers; it has been known for many years that their mechanical properties can

approach those which are theoretically possible. This is because of the
lack of defects which weaken other forms of matter. The requirement
that the whiskers be free of defects usually means that they have to be
extremely small with diameters of around 1 μm and less and lengths
usually not more than a few millimetres. Larger dimensions can result in
dramatic falls in mechanical properties as is exemplified in Figure 2.18
for potassium titanate whiskers [34]. Despite their small dimensions
whiskers have potential as reinforcements because of their high aspect
ratios. Whiskers can be produced from many crystalline materials; some
relevant examples are given in Table 2.4. Examples of commercially
available whiskers are also given in chapter 9. Methods of producing
whiskers can be exemplified with SiC whiskers which are the most
widely available commercially. Six production routes can be distin-
guished [35]:

(1) the thermal decomposition or hydrogen reduction of organic
    silicon compounds such as $Si(CH_3)_3Cl$ or $SiCH_3Cl_3$ in the tem-
    perature range 1100–1500 °C;
(2) hydrogen reduction of a gaseous mixture of silicon halides such as
    $SiCl_4$ and hydrocarbons or CCl_4 in the temperature range
    1200–1500 °C;
(3) recrystallisation of sublimed silicon carbide around 2500 °C;
(4) supersaturation technique in the molten phase of silicon alloys at
    high temperature using Fe, Ni or NaF as an agent;

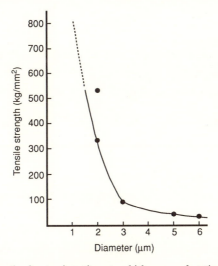

Figure 2.18 The strength of potassium titanate whiskers as a function of diameter (Otsuka
Chemical data [34]).

Table 2.4　Comparison of whisker types and properties

Whisker	Tensile strength (GPa)	Elastic modulus (GPa)	Density (g/cm^3)	Melting or softening temperature (°C)
Alumina	20	450	3.96	2040
Beryllia	13	350	2.85	2570
B$_4$C	14	490	2.52	2450
Graphite	19	700	2.00	13000
SiC	21	490	3.18	2690
Si$_3$N$_4$	14	385	3.18	1900
Potassium titanate	7	280	3.3	1300

(5) reaction between SiO_2 and C with a catalyst in the temperature range 1300–1700 °C;

(6) reaction between Si and a hydrocarbon such as C_3H_8 in the presence of H_2S, which produces an intermediate compound of SiS or SiS_2, in the temperature range 1200–1300 °C.

Processes 1 and 2 are the most popular CVD techniques for producing pure SiC whiskers. Processes 3 and 4 are convenient methods for growing large size single crystals. Process 5 is used for the mass production of SiC whiskers with various types of raw materials and catalysts. Process 6 enables the easy preparation of whiskers with lengths of several millimetres and diameters of less than 1 μm.

Two mechanisms have been proposed to explain the growth of SiC whiskers. One is growth from the vapour phase by screw dislocation along the axial direction, the other is by which the whisker deposits from a supersaturated solution of Si and C in a liquid drop including a catalyst such as iron [36]. The mechanical properties of whiskers are limited by defects. These can be stacking faults and twin dislocations on close packed planes lying normal to the whisker axis. Cavities 1–20 nm in diameter near the whisker core with partial dislocations lying in radial directions have also been observed. Further details of whisker properties are given in chapter 9.

2.6　Conclusions

A range of small diameter ceramic fibres and whiskers have been developed that have some potential as reinforcements in ceramic matrices. The fibres that are currently produced in processes involving organo-metallic precursor fibres are rarely of simple composition. One consequence of this is that they often have lower elastic moduli and higher strains to failure than pure ceramics. The latter characteristic is beneficial in ceramic and glass-matrix composites since it is one prerequisite of effective fibre toughening. However, another consequence is that the fibres exhibit considerable weakening above about 1000 °C. Exceptions to this are carbon fibres which remain stable to very high temperatures but only if protected from oxidation. Considerable effort is currently being made to develop ceramic fibres that retain their properties between 1000 and 2000 °C. Success in this would be an important step in the development of ceramic composites.

References

1. Wawner, F. E. (1988) Boron and silicon carbide/carbon fibres in: *Fibre Reinforcements for Composites Materials*, ed. Bunsell, A. R., Elsevier, Amsterdam, p. 371.
2. Di Carlo, J. A. (1984) High performance fibers for structurally reliable metal and ceramic composites, NASA Tech. Mem 86878.
3. Champion, A. R., Kreuger, W. H., Hartman, H. S. and Dhingra, A. K. (1978) *Proc. ICCM 2*, eds. Moton, B., Signorelli, R., Street, K. and Phillips, L., TMS AIME, New York, p. 883.
4. Yajima, S., Okamura, K., Hayashi, J. and Omori, M. (1976) Synthesis of continuous SiC fibers with high tensile strength, *J. Am. Ceram. Soc.* 59, 324.
5. Fitzer, E. and Heine, M. (1988) Carbon fibre manufacture and surface treatment in: *Fibre Reinforcements for Composite Materials*, ed. Bunsell, A. R., Elsevier, Amsterdam, p. 73.
6. Shaffer, P. T. B. (1967) Whiskers. Their growth and properties, in: *Modern Composite Materials*, eds. Broutman, L. J. and Krock, R. H., Addison-Wesley, Reading, MA, p. 197.
7. Dhingra, A. K. (1980) Alumina fibre FP, *Philos. Trans. R. Soc. London* A294, 411.
8. Shin, H. (1975) Process for preparing alumina yarns, U.K. Patent N° 1414854.
9. Romine, J. C. (1987) New high temperature ceramic fiber, *Ceram. Eng. Sci. Proc.* 8, 755.
10. Dhingra, A. K. (1984) Advances in inorganic fiber developments, in: *Contemporary Topics in Polymer Science*, Vol. 5, ed. Vandenberg, E. J., Plenum Press, New York, p. 227.
11. Abe, Y., Horikiri, S., Fujimura, R. and Ichiki, E. (1982) in *Progress in Science and Engineering of Composites, Proc. ICCM-V*, ed. Hayashi, T., Tokyo, p. 1427.
12. Bunsell, A. R. (1988) Development of fine ceramic fibres for high temperature composites, *Materials Forum* 11, 78.
13. Cannon, W. R. and Landgon, T. G. (1983) Review: creep of ceramics, *J. Mater. Sci.* 18, 1.
14. Bunsell, A. R., Simon, G., Abe, Y. and Akiyama, M. (1988) Ceramic fibers, in: *Fibre Reinforcements for Composite Materials*, ed. Bunsell, A. R., Elsevier, Amsterdam, p. 427.
15. Symes, W. R. and Rasteller, E. (1981) Alumina fibre, in: *Proc. 24th Int. Coll. on Refractories*, Aachen.
16. Yajima, S., Hasegawa, X., Hayashi, J. and Iimua, M. (1978) Synthesis of continuous silicon carbide fibre with high tensile strength and high Young's modulus, *J. Mater. Sci.* 13, 2569.
17. Hasegawa, Y., Iimura, M. and Yajima, S. (1980) Synthesis of continuous silicon carbide fibre Part II, Conversion of polycarbosilane fibre into silicon carbide fibres, *J. Mater. Sci.* **15**, 720.

18. Simon, G. and Bunsell, A. R. (1984) Mechanical and structural characterisation of the Nicalon carbide fibre, *J. Mater. Sci.* 19, 3649.
19. Simon, G. and Bunsell, A. R. (1984) Creep behaviour and structural characterisation at high temperatures of Nicalon SiC fibres, *J. Mater. Sci.* 9, 3658.
20. Frety, N. and Boussuge, M. (1991) Relationships between the high temperature evolution of different fibre matrix interfaces and the mechanical behaviour of SiC-SiC composites, *Composite Sci. Tech.*, to be published.
21. Favry, Y. and Bunsell, A. R. (1987) Characterisation of Nicalon (SiC) reinforced aluminium wire as a function of temperature, *Composite Sci. Tech.* 30, 85.
22. Moore, G. A. (1991) Ceramic fibres, to be published.
23. Yamamura, T., Hurushima, T., Kimoto, M., Ishikawa, T., Shibuya, M. and Wai, T. I. (1987) Development of new continuous Si-Ti-C-O fibre with high mechanical strength and heat resistance, in: *High Tech. Ceramics*, ed. Vincenzini, P., Elsevier, Amsterdam, p. 737.
24. Penn, B. G., Ledbetter III, F. E., Clemons, J. M. and Daniels, J. G. (1982) Preparation of silicon carbide-silicon nitride fibres by the controlled pyrolysis of polycarbosilazane precursors, *J. Appl. Polymer Sci.* 27, 3751.
25. Ledbetter III, F. E., Daniels, J. G., Clemons, J. M., Hundley, N. H. and Penn, B. G. (1984) Thermogravimetric analysis of silicon carbide-silicon nitride polycarbosilazane precursor during pyrolysis from ambient to 1000 °C, *J. Mater. Sci.* 19, 802.
26. Lipowitz, J., Freeman, H. A., Chen, R. T and Prack, E. R. (1987) Composition and structure of ceramic fibres prepared from polymer precursors, *Adv. Ceram. Mater.* 2, 121.
27. Tai-Il Mah, Mendiratta, M. G., Katz, A. P. and Mazdiyasni, K. S. (1987) Recent developments in fibre reinforced high temperature ceramic composites, *Ceram. Bull.* 66, 304.
28. Sawyer, L. C., Jamisson, M., Brikawski, D., Haider, M. I. and Chen, R. T. (1987) Strength structure and fracture properties of ceramic fibres produced from polymeric precursors, *J. Am. Ceram. Soc.* 70, 798.
29. Salinger, R. M., Barnard, T. D., Li, C. T. and Mahone, L. G. (1988) Utilization of polymer precursors in the formations of Si-N-C advanced ceramic fibres, *SAMPE Quarterly* April, 27.
30. Gourdin, Ch., private communication.
31. Fitzer, E. and Herise, M. (1988) Carbon fibre manufacture and surface treatment, in: *Fibre Reinforcements for Composite Materials*, ed. Bunsell, A. R., Elsevier, Amsterdam, p. 73.
32. Oberlin, A. and Guigon, M. (1988) The structure of carbon fibres, in: *Fibre Reinforcements for Composite Materials*, ed. Bunsell, A. R., Elsevier, Amsterdam, p. 149.
33. Sambell, R. A. J., Phillips, D. C. and Bowen, D. H. (1974) The technology of carbon fibre reinforced glasses and ceramics, UKA.EA Report AERE-R-7612; in: *Proc. Carbon Fibre: Their Place in Modern Technology*, PRI, p. 105.
34. Data from Otsuka Chemical Co. Ltd., Osaka (1988).
35. Bracke, P., Schurmans, H. and Verhoest, J. (1984) *Inorganic Fibres and Composite Materials*, Pergamon Press, Oxford, pp. 61–75.
36. Milewski, J. V., Gac, F. D., Petrovic, J. J. and Skaggs, S. R. (1985) *J. Mater. Sci.* 20, 1160–1166.

3 Principles of preparation of ceramic composites

R. WARREN and R. LUNDBERG

3.1 Introduction

As already indicated in chapter 1, the goal of production processes for
ceramic composites is to produce materials with a predetermined microstructural geometry with a minimum of harmful defects. For the
preparation of particulate and whisker reinforced composites it has been
found possible to draw largely on processing techniques developed for
traditional and monolithic ceramics. For long-fibre composites, however,
special new techniques have had to be developed. This is partly because
continuous fibre geometries (usually multiaxial) have to be preserved
during processing with a minimum of fibre damage and partly because
many fibres are degraded at the temperatures required for the sintering
of ceramics. Such degradation can be the result of structural change in
the fibre (see chapter 2) or chemical reaction with the matrix. An
underlying principle of techniques developed for long-fibre composites is
that a preform of fibres with the required geometry is infiltrated with
the matrix or a marix precursor. The infiltrating matrix can for example
take the form of a powder slurry (slurry infiltration), a melt (melt
infiltration), a liquid solution (liquid precursor infiltration) or a mixture
of gases or vapours that react *in situ* to form the matrix (chemical
vapour infiltration, CVI).

A factor that has to be considered in all composite processing is the
need to be able to optimise the strength of the fibre/matrix interface
since this has a critical influence on properties (see chapter 4).

The present chapter provides an introduction to the principles underlying the processing methods for all the above types of composite;
examples are of course detailed in the later chapters devoted to specific
composites. As a basis for descriptions of the processing of particulate
composites as well as to permit an appreciation of the difficulties
encountered with whisker and long-fibre composites, the chapter begins
with an outline of processing methods used for conventional ceramics.

3.2 Processing of monolithic ceramics and particulate composites

3.2.1 *Basic stages*

The basic steps in the preparation of ceramic components by established powder processing are (see Figure 3.1):

- preparation of powder
- milling and/or mixing of constituents and additives
- forming of green compact
- sintering

Needless to say, in practice, a process normally comprises several other intermediate secondary steps such as drying, removal of additives, etc. Moreover, a number of techniques have been developed that combine two of the above stages, e.g. co-precipitation of powders to combine the first two stages or uniaxial hot pressing (HP) which combines the last two. A number of alternative processing routes are included in abbreviated form in Figure 3.1. More detailed examples for particulate and whisker composites are given in chapters 6 and 9.

The increasing interest in ceramics as structural and functional materials and, in particular, the need to improve their mechanical reliability has led to substantial developments in processing as well as in its underlying principles. For example, studies of sintering have led to the realisation not only that sinterability is promoted by small powder particle size and constituent purity but also that the homogeneity and

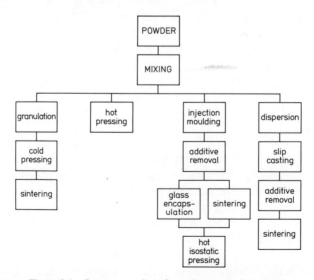

Figure 3.1 Some examples of powder processing routes.

size distribution of residual porosity after sintering are sensitive to the powder particle size distribution and to local density variations in the green compact. Thus the population of large, harmful sintering defects can be minimised by use of powders with a controlled, narrow particle size distribution and by the avoidance of agglomerates. This has been achieved by developments in powder preparation and control of powder dispersion through understanding of the physical chemistry of powder surfaces. Other developments have included the improvement of forming and sintering processes. Thus, improvements in density have been achieved both by the development of combined pressure/sintering processes and by the use of sintering aids. Various reactive sintering processes have also been developed with some success. Of considerable practical importance has been the development of powder forming techniques which in combination with appropriate sintering methods lead to net or near-net shaped components obviating the need for expensive and harmful post-sinter machining operations.

These points are described in more detail in the following sections.

3.2.2 *Preparation of powders*

The principle source of the powder constituents of ceramics are naturally occurring minerals which are refined via a series of mechanical and chemical extraction stages to yield the required product [1]. For economic production of large volumes, the ceramic is produced with as few steps as possible, preferably in continuous or large scale batch processes. However, for more advanced ceramics requiring improved control of impurity levels and particle morphology, more involved chemical processes have been developed. These can be broadly divided into those based on *precipitation from solution* and those based on *vapour phase reactions*.

3.2.2.1 Large scale processes. Examples of larger scale processes for the powders of the principal monolithic ceramics are as follows:

Alumina is prepared from the mineral, bauxite, using the Bayer process [1, 2] which involves taking the mineral into solution as sodium aluminate by reaction with NaOH. After filtering, aluminium hydroxide is precipitated. This is then calcined at 1100–1200 °C to produce alumina.

Zirconia is prepared from the mineral zircon ($ZrSiO_4$) [2]. For example, by reaction with NaOH above 600 °C sodium zirconate and sodium silicate are formed. Leaching out the latter with water leaves a hydrated zirconium hydroxide which can be calcined directly to zirconia of moderate purity. Higher purity can be attained by reacting the hydroxide with sulphuric acid to give zirconium sulphate which can be calcined.

Silicon carbide is prepared by carbothermic reduction either of naturally occurring quartz or of silica derived from other silicon-containing minerals [2, 3]. In the well-known Acheson process the source of carbon is coke or coal and the reaction is carried out in excess of 2600 °C in an electric resistance furnace. The product is in the form of coarse α-SiC crystals. β-SiC can be formed in lower temperature processes but this normally requires the presence of catalysts.

Silicon nitride is commonly prepared by nitridation of silicon (prepared from silica) at around 1400 °C to produce predominantly α-Si$_3$N$_4$ [2, 4] (the alpha form is preferred in the preparation of silicon nitride ceramics; during sintering the alpha form transforms to the beta form with more favourable microstructure and properties [5]). Another important process is the carbothermic reduction of silica in the presence of nitrogen [4]:

$$3SiO_2 + 6C \xrightarrow{2N_2} Si_3N_4 + 6CO \qquad (3.1)$$

In their basic forms the processes described above give a relatively low-cost product but this is in the form of relatively impure coarse agglomerates that normally have to be milled (see below).

3.2.2.2 *Methods based on solutions and precipitation from solutions.*

An underlying approach common to many processes based on precipitation from solution is to form a solution of a salt or salts from which hydroxides or hydrated oxides can be precipitated and subsequently calcined to form oxide powders. The dissolved salts can be either inorganic, e.g. nitrates, chlorides etc., or organic, e.g. acetates or metal complexes such as alkoxides. Precipitation can be achieved for example by seeding a saturated solution or chemically by the addition of an appropriate reagent such as ammonium hydroxide. A very fruitful technique is to create the precipitate in the form of a *sol* (i.e. a suspension of extremely fine particles or large molecules). This can be further transformed to a gel (i.e. the sol particles link up to form a loose skeletal structure) which can be dried, crushed and calcined. The particles in the *gel* are generally very fine as well as being loosely bound. Consequently, it can usually be reduced to a fine powder by relatively mild milling and calcined at relatively low temperatures.

Further developments of solution processes involve the spraying of solutions or sols, e.g. spray drying, freeze drying [6], spray pyrolysis (i.e. spraying into a high-temperature chamber [7]) and electrostatic atomisation [8]. Each drop in the spray dries to leave a fine particle or agglomerate. An advantage of spray processes is that in preparing a mixed powder (e.g. a composite powder) the risk of separation of the constituents is eliminated. A process related to spraying involves forming an emulsion of solution or sol droplets in a second immiscible liquid. The droplets can be gelled to provide particles of desired size.

By modification of the variables in such processes, a wide range of particle morphologies can be achieved, e.g. dense, submicron spheres with uniform size, large agglomerated spheres (to give a powder with good flow characteristics) and fibrous particles. Mixed powders suitable for particulate composites can be prepared by producing mixed sols or sol/solution combinations. Similarly, solid solution oxides and glasses can be prepared [9]. It can also be noted here that sol-gel methods can be exploited to introduce matrix constituents into a composite by infiltration of a fibre preform with a sol, the subsequent gelling and calcining taking place *in situ* (see section 4).

The literature on solution and sol-gel techniques is extensive and cannot be treated exhaustively here. Three recent reviews are those of Segal [10], Jones [11] and Brinker and Scherer [12]. An appropriate, illustrative example is the preparation of an alumina/zirconia mixed powder [13]. Hydrated aluminium chloride ($AlCl_3.6H_2O$) and zirconium chloride ($ZrCl_4$) were dissolved in distilled water in proportions corresponding to 10 wt% ZrO_2 in the final powder. Hydroxides of the two metals were precipitated as a gel by adding NH_4OH and stirring until a pH of 9 was reached. The gel was washed and dried and then redispersed as a sol in an HCl solution to give a concentration of 4 mol/l Al^{3+} ions. This sol was broken up into droplets in the organic liquid 1-octanol which dehydrated the sol to give solid gel microspheres of hydroxide. The octanol contained 1% of a surfactant (see below) to prevent agglomeration of the spheres. These were calcined, e.g. at 1200 °C for 30 min, to give mixed oxide microspheres with an average diameter of 20 μm.

Although the majority of solution and sol-gel methods have been developed for the preparation of oxide powders, such processing routes can also be devised for non-oxide ceramics. For example, silicon carbide and silicon nitride powders can be prepared via solutions of silicon-containing organic compounds such as silanes with an ultimate stage in which an organic polymer is pyrolysed [14]. Alternatively, the carbothermic reactions described earlier can be applied to ultrafine silica particles produced by a solution-based process. Analogous processes are feasible for boron carbide and boron nitride powders starting with boron-containing compounds. In the pyrolysis of the organic compounds to yield non-oxide ceramics it is usually difficult to achieve exact stoichiometry and complete elimination of oxides.

Synthesis processes based on sol-gel methods for oxides and polymer pyrolysis for non-oxides are also used in the preparation of continuous fibres; several examples are given in chapter 2.

3.2.2.3 Processes based on vapour phase reactions. Two important groups of processes based on vapour phase reactions are *chemical vapour reaction* and *plasma assisted reaction*. Almost any ceramic compound can be formed by reaction between suitable vapours and gases, usually metal chlorides together with carbon-, nitrogen- or oxygen-containing gases as appropriate. A comprehensive list of relevant reactions has been given by Stinton *et al.* [15]. Important selected examples are shown in Table 3.1.

Typical reaction temperatures are between 900 and 1200 °C. The reactions are identical to those used for chemical vapour deposition of coatings and the chemical infiltration of fibre preforms (see section 3.4.2 and chapter 8). A powder reaction product rather than a coating is achieved by ensuring the absence of a hot substrate. Similarly the morphology of the powder can be adjusted through the process parameters (temperature, partial pressure of reactants, etc.). A disadvantage is that only single phase powders can be produced at one time [16].

In plasma synthesis, similar reactions to the above are conducted in arcs formed between electrodes, either at low or atmospheric pressures. Extremely fine particles with diameters down to tens of nanometers are achievable [17]. Plasma methods are amenable to scaling up to relatively high production rates and a number of powders are now produced commercially.

3.2.3 Milling, mixing and dispersing

3.2.3.1 Milling. Powders produced in processes yielding coarse particles or agglomerates of fine particles must normally be reduced to appropriate particle size in powder mills. A common feature of the various types of mill is that they consist of a closed chamber in which the powder is placed together with milling bodies (e.g. balls or cylinders) which when the mill is in motion exert a crushing action on the powder [1]. The milling action is most effective if the milling bodies have high density. However to avoid contamination of the powder, ceramics are often milled with ceramic bodies which prolongs the necessary milling time. Some tens of hours are usually necessary to

Table 3.1 Examples of vapour phase reactions that give ceramic reaction products (taken from [15])

$SiCl_4 + NH_3 + H_2$	give	Si_3N_4
$CH_3SiCl_3 + H_2$	give	SiC
$AlCl_3 + CO_2 + H_2$	give	Al_2O_3
$ZrCl_4 + CO_2 + H_2$	give	ZrO_2
$TiCl_4 + BCl_3 + H_2$	give	TiB_2

reduce the particle size to around 1 μm. When milling is necessary then it can be exploited to provide an effective mixing and dispersion of the powder constituents of particulate composites and to mix in other processing additives where required (see below). Milling is not normally suitable for mixing whisker or short-fibre composite constituents since the fibres are quickly destroyed.

3.2.3.2 Mixing and dispersing. Provided the particle size is sufficiently small, mixing of powder constituents and additives can be achieved by dispersing them in a liquid using a mixer. These generally consist of rotating units that create high shearing forces in the liquid [1]. They can be so designed and adjusted that they break up agglomerates without necessarily destroying short fibres. Ultrasonic agitation also provides effective dispersion in many cases.

As well as mixing the constituents of the material itself (i.e. sintering aids as well as the constituent phases (see section 3.2.5)) various processing additives required in later forming processes must be included as appropriate. Examples are lubricants for cold pressing, binders to hold together green compacts of the powder, carriers for powders to be injection moulded or extruded, and surfactants to maintain the constituent powders in suspension. In certain cases, the additives can be taken into solution in the liquid medium or adsorbed at the particle/liquid interface. Examples of such additives are given below in descriptions of various forming processes.

Mixing of powder constituents can be achieved very effectively if the powders are fine and can be kept in stable suspension. This is usually also a necessary condition for maintaining a mixture, i.e. to prevent separation by gravitational settling or by flocculation (i.e. separation into weakly-bound aggregates). Moreover, a stable suspension is necessary in creating slurries, e.g. for slip casting (section 3.2.4). A stable suspension of particles is achieved if they are fine and if a repulsive force is set up between them to counterbalance attractive van der Waal forces [1, 18]. In aqueous solutions such a repulsive force can be created through the formation of an electric double layer on the particle surfaces, so-called electrostatic stabilisation. This occurs in a range of pH that is characteristic for each powder. This can lead to difficulties in stabilisation of powder mixtures since the stabilising pH range of the constituents may differ. Bostedt *et al.* [18] have shown that to circumvent this problem particles of various compounds can be given an alumina-like surface by treatment with Al-alkoxide and thereby achieve similar electrostatic dispersion behaviour. A suitable repulsive force can also be achieved by the adsorption of certain organic polymers at the powder/liquid interface, so-called steric stabilisation. Such steric stabilisers can be found for both aqueous and organic liquid suspensions and can function with the different constituents of a mixture. For environmental reasons aqueous suspensions are to be preferred.

Once a homogeneous mixture has been achieved by such dispersion techniques it is possible to 'fix' the mixture, for example by a spray technique or by flocculation (i.e. separation into weakly bound agglomerates). The floc itself can be mixed further to produce a more homogeneous mixture that remains very stable [19]. In certain forming process such as slip casting (see section 3.2.4.3) a compact is formed direct from the suspension. In such cases it is important that the constituents do not separate during the forming process.

3.2.4   *Powder forming processes*

The purpose of a powder forming process is to shape the powder as nearly as possible into the shape of the intended component and at the same time bring the powder particles as closely together as possible to permit effective sintering. In cold forming processes the powder is formed into a green compact prior to subsequent sintering while in hot forming processes forming and sintering occur simultaneously. The effectiveness of sintering of green compacts is favoured by high green density and by uniformity of the distribution of the powder particles (i.e. a minimum of local density variations caused for example by cavities and agglomerates). Similarly the control and retention of shape during sintering are also favoured by a high green density which implies low shrinkage and a controllable variation of green density within the compact. These considerations underly the development of cold forming processes and the nature of the powders used in them.

The principal cold forming processes discussed here, namely uniaxial pressing, cold isostatic pressing, slip casting, injection moulding and extrusion are illustrated in Figure 3.2 (hot forming processes are treated in the next section concerned with sintering). The choice of method is determined by the geometry, the complexity of shape and the production volume of the component.

3.2.4.1   Uniaxial pressing.   In uniaxial cold pressing, the powder is poured into a rigid die and is then pressed with a closely fitting punch [1]. After pressing the punch is removed and the green compact is removed, usually by ejection with a lower punch. An overriding concern is that friction between powder particles and between powder and die/punches lead to an uneven distribution of stress throughout the compact. This leads not only to non-uniformity of density but also to variation of elastic relaxation upon unloading. The latter effect can lead to fracture of the compact and this limits the pressures that can be used

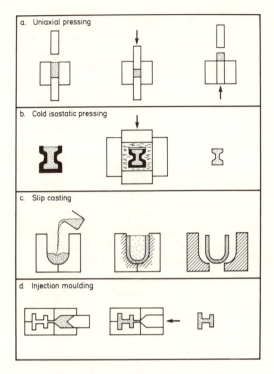

Figure 3.2 Principal methods of powder forming (a) uniaxial pressing, (b) cold isostatic pressing, (c) slip casting, (d) injection moulding.

to about 50 MPa giving green densities of 50–60% of full density. The friction effects also limit the size and geometry of the compact. For example the ratio of axial height to width of the compact cannot normally exceed 3. To minimise friction a small amount of lubricant is added to the powder, typical examples being stearic acid and metal stearates. Adsorbed polymers can also be effective lubricants, being present as a thin layer on all particles. In addition to lubricants, binders are added to provide binding between particles to impart green strength to the compact sufficient for it to survive ejection and subsequent handling. Sometimes, the compacts are given sufficient green strength to permit machining to more complicated shapes. Typical binders are polyvinyl alcohol in combination with polyethylene glycol as plasticiser and lignosulphonates. Up to about 10% of such additives are employed and must be removed by controlled heating at a later stage (see below).

Another important practical consideration is the powder flowability which determines the ease with which the die can be filled, a critical factor in automated, industrial production. Powders with micrometre sized particles do not flow easily and will not fill a die uniformly. For this reason larger, loosely bound, spherical agglomerates or granules are prepared either at the powder preparation stage (see section 3.2.2) or by spray drying a slurry of the powder with appropriate additives. Clearly the granules have to be easily deformed during pressing to avoid non-uniformity of density in the green compact. An alternative is to fill the die directly with slurry and arrange for an effective removal of liquid during pressing [20].

Uniaxial pressing becomes difficult if the powder contains whiskers or short fibres. The random packing density of particles with fibrous shape is very poor (section 3.3 and Figure 3.3). If the fibres are not to become crushed, the achievable green density is low. Moreover, elastic spring-back effects are large.

In cold uniaxial pressing the shape and size of the compact is not only limited by the friction effects described above but also by the fact that it has to be ejected from the die. Thus the method is suitable for small components with relatively simple geometry. Further, the high investment cost of press and tooling together with the potential for automation mean that the method is best suited for production of large series.

3.2.4.2 Cold isostatic pressing. In this process the powder is filled into a flexible mould generally prepared from rubber [1]. This is then subjected to an isostatic pressure via a fluid in a pressure chamber. Friction effects are much less in this process than in uniaxial pressing and the pressing of larger more complex shapes is possible. As in uniaxial pressing, lubricants and binders are used and green machining is sometimes employed. The rubber moulds are relatively easy to produce but the process cycle itself is slow and so the method is appropriate for small series of components.

3.2.4.3 Slip casting. The principle of slip casting is that a slurry of the powder is poured into a porous mould, usually made of gypsum, which absorbs the liquid carrier causing the powder to be drawn to the mould walls [1]. The method is well known for the production of traditional ceramics and is ideal for the preparation of large, thin-walled, hollow components. Recent developments in powder preparation and colloid science have extended its use to advanced ceramics. For optimum casting the slurry or slip should be a stable suspension with as high a solid content as possible but with low viscosity. By suitable choice of stabilisers, stable slips of most ceramics and mixtures of ceramics can be achieved in aqueous as well as organic liquids (see previous section). Particle sizes lie between 0.1 and 1 μm; particles smaller than 0.1 μm become increasingly difficult to stabilise and consolidate, thus requiring increasing amounts of additives. Examples of stabilisers in aqueous slips are:

(1) for alumina: ammonium polyacrylate [1]; polyacrylic acid
                          [21]; sodium carboxymethylcellulose [22].
(2) for silicon carbide: polyethyleneimines [23].

Solid contents of up to about 40 vol% and green densities of up to 70 vol% have been achieved. Depending on the thickness of the deposit, casting can take up to several hours but the time can be reduced by application of pressure to the slip or of vacuum to the mould.

A process related to slip casting is *tape casting* in which thin layers of slurry (0.01–1 mm) are cast onto a substrate of plastic film.

Both tape casting and slip casting are suitable for forming short fibre composites, tending to produce a two-dimensionally random fibre orientation [24].

3.2.4.4 Injection moulding. Injection moulding [1, 25] as applied to ceramics involves mixing the ceramic powder with a sufficient amount of polymer or other soft binder to produce a mouldable dough. This is then injected under pressure into a mould. With appropriate flow properties and strength, the mix can be moulded to complex and intricate shapes. The mould is constructed of demountable parts, the design of which is dictated by the complexity of the ceramic component.

The binders are generally mixtures consisting most commonly of thermoplastic polymers such as high molecular weight polystyrene, polyethylene and various waxes. Thus the mixing of the binder with the powder (usually as granules) as well as the moulding can be carried out at temperatures above the glass transition where the polymer is soft. On cooling, the moulded component becomes deformation resistant and handleable. The mix also contains dispersing agents and lubricant etc. Typical injection pressures are between 1 and 100 MPa depending on the binder system chosen. After moulding and prior to sintering the binder has to be extracted or burnt off; one reason for using mixed binders is to give a successive and gradual binder removal (see next section).

It is clearly desirable to produce mixes with as high ceramic powder content as possible in order to minimise the subsequent binder removal that is necessary and to achieve a high green density. The powder content is however limited to a critical volume fraction above which the viscosity of the mix increases sharply. The limit is set approximately by the point at which the particles begin to develop skeletal contact (i.e. it is related to the percolation limit discussed in chapter 4 in connection with conductivity) and is therefore very sensitive to particle shape and particle size distribution. A relevant example is the case of fibre-shaped particles; for randomly oriented fibres the unconstrained packing density falls from about 65 vol% to less than 10% when the length/diameter ratio increases from 1 to 50 (Figure 3.3) [25, 26]. The packing density limit corresponds to the onset of a locked skeleton. Application of pressure would lead first to elastic springback on unloading and then fibre fracture. To be readily deformable the ceramic content would need to be a little below the packing limit. Spherical powders permit dense packing and up to 60 vol% in the moulding mix is possible.

Injection moulding has potential for the fabrication of short-fibre reinforced composites; of particular interest is the possibility of controlling fibre orientation through control of the flow of the mix into and within the mould [25]. Preliminary studies of the injection moulding of fibre composites have been reported; a problem is the limit to fibre fraction set by packing geometry [25].

Injection moulding is suited to the production of moderately large series of components with complex shape. A related process, namely *extrusion*, is suitable for producing rod and strip with constant cross-sections.

3.2.4.5 Drying, additive removal and presintering. From the above accounts of powder processing it is clear that the removal of various additives from powder compacts is an important step prior to sintering. This is generally achieved by heating to cause either evaporation or oxidation (i.e. burning off). For processes such as injection moulding in which the compact contains as much as 50 vol% additives, a large proportion can first be removed by dissolution in a suitable solvent or by wicking, that is extraction of the molten binder into a porous bed surrounding the compact. Wicking leaves enough binder wetting the particles to hold the compact together for subsequent handling. With a mixed binder system the constituent with the lowest melting point is

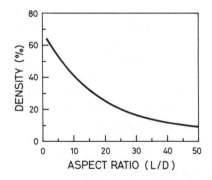

Figure 3.3   Effect of fibre aspect ratio on the packing density of randomly oriented fibres (after [26]).

extracted first leaving channels for removal of the remaining constituents. The theory and practice of additive removal using both liquid extraction and evaporation methods are reviewed well by German [25].

An important feature of all extraction processes is that the additives must leave the compact via the pore system. The process begins at the surface and the interior material must dissolve or evaporate and migrate to the surface. Heating rates and holding times must be carefully controlled to avoid cracking and distortion of the compact due to overrapid evolution. In most extraction systems the process is monitored continuously (e.g. by weight measurement or gas pressure) and the measured parameter used to optimise the heating via a suitable computer control programme. Even when optimised in this way the process can take several hours.

When the additive removal is complete there is a risk that the compact will be too fragile to handle. Thus the final stage of removal may be arranged to take place during sintering or in a presintering process in which a sufficiently high temperature is reached to provide some sintering between particles. Presintering can also be exploited to produce a compact that can be readily machined to complex shapes.

3.2.5 *Sintering processes*

The basic process for consolidating ceramic compacts is pressureless, solid state sintering. A multitude of modifications to this process such as pressure assisted sintering, liquid phase sintering and reaction bonding, have been developed to accelerate the densification and above all to reduce the necessary temperature. The principles of solid state sintering and some of the modifications will be outlined briefly below.

3.2.5.1 Solid state sintering. The two main functions of sintering of powder compacts are (a) the development of atomic bonding between particles to provide strength and (b) the elimination of porosity. The driving force for these processes is provided by the consequent replacement of the solid surface energy of the particles by grain boundary energy (for complete densification the grain boundary area is half the original particle surface area).

The main stages in solid state sintering are illustrated in Figure 3.4. Firstly, mechanical contacts between particles grow into necks across which atomic bonding is developed, thus creating grain boundaries (Figure 3.4b). This can occur in association with smoothing of the particle surfaces by local surface diffusion which can occur at relatively low temperatures and short times and without significant shrinkage. Thus necks can form and pore shapes change without a reduction in

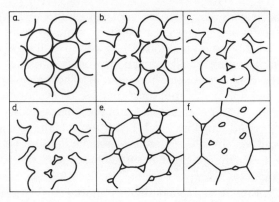

Figure 3.4 Stages of solid state sintering (a) green compact, (b) neck formation, (c) rearrangement, (d) open pore stage, (e) closed pore stage, (f) break-away.

pore volume. Some shrinkage can be achieved during this early stage by rearrangement of particles [27, 28]. This can occur by local diffusion around particle contacts which can lead to relative particle motion until all the particles become locked in a three-dimensional skeleton (Figure 3.4c). As already indicated, green densities are usually around 50–60% and a skeletal structure develops rapidly, leading to the second main stage of sintering in which the pores remain as an interconnected network open to the surface of the compact (Figure 3.4d). Densification implies that there is an overall movement of atoms from the compact surface to the interior which can occur readily at this stage by volume, grain-boundary and surface diffusion processes. When the compact reaches about 90% density, the open pore structure begins to break up into isolated pores held predominantly at the grain boundaries (Figure 3.4e). Continued densification can occur by grain-boundary diffusion but now any gas trapped in the pores must also diffuse out which requires that it is soluble in the ceramic.

During sintering, grain growth also occurs. This is undesirable, firstly because it reduces the rate of sintering by reducing the grain-boundary density and secondly because it usually affects mechanical properties negatively. The driving force for grain growth is a reduction in grain-boundary energy and the rate of growth is set by the boundary mobility (i.e. the movement of atoms across the boundary). Both the driving force and the mobility are reduced by pores in the boundary and consequently grain growth is slow during early stages of sintering. However, when the volume fraction porosity is sufficiently low and the grain size sufficiently large the grain boundaries can break away from the pores leaving them isolated in the grains (Figure 3.4f). In this stage, densification can only occur by volume diffusion and becomes very limited; at the same time grain growth rate increases significantly.

Against this background, the main features of sintering outlined in Figure 3.5 can be understood. For isothermal sintering (Figure 3.5a), the sintering rate measured in terms of, for example, increase in density, is initially high (in practice, a significant proportion of the early stage shrinkage occurs during heating to the sintering temperature) but falls rapidly as shrinkage mechanisms become exhausted. The sintering rate and density achieved after a given time increase with temperature and decreasing grain size. With increasing initial density in the green compact, the shrinkage rate and total shrinkage decrease but a higher density is attained for a given sintering time. This effect can cause fissure formation, non-uniform shrinkage and distortion in compacts with non-uniform green density.

Shrinkage rate normally increases exponentially with temperature (Figure 3.5b) since the underlying diffusion processes are thermally activated. For ceramics, sintering temperatures lie typically in the range 1500–2200 °C. For most, particularly those with mainly covalent bonding (e.g. silicon nitride and silicon carbide), the diffusion rates are low and sintering to high density is difficult. Effective densification cannot necessarily be achieved simply by increasing temperature since densification is eventually restricted by evaporation of the ceramic or by excessive grain growth. For composites there is also the likelihood of reaction between the constituents or fibre degradation. Evaporation can be reduced to some extent by embedding the compact in a powder bed or by using a high pressure sintering atmosphere. The interrelationship between grain growth and densification can be represented in a plot of grain size versus density for a given sintering situation (Figure 3.5c). As the compact approaches full density the grain growth accelerates. In certain cases the breakaway grain growth conditions will be reached. A goal of sintering modifications is to shift the grain size/density curve as far to the bottom right of the diagram as possible. This can be achieved for instance by the use of sintering additives (see below).

In many composites the addition of the second phase has the desired effect of increasing the ratio of densification rate to grain growth rate.

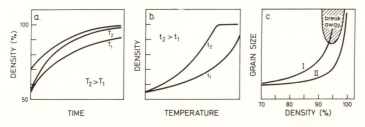

Figure 3.5 Schematic illustration of solid state sintering (a) isothermal densification curves, (b) isochronal densification temperature diagram, (c) grain size/densification map (curve II represents improved grain growth inhibition).

Thus second phase particles are very effective in reducing grain growth by boundary pinning [29] while the densification rate of a mixture of particulate phases usually lies between that of the pure constituents. The situation is less favourable with short fibres and whiskers. The fibres generally have poor intrinsic sinterability (indeed, they are required to maintain their shape during sintering) and they therefore inhibit the shrinkage of the surrounding matrix particles. Their shape also makes them less effective in grain-boundary pinning. As an example, additions of 0–15 vol% SiC whiskers in alumina, sintered without pressure at 1500 °C caused a proportional reduction in density from 96% to 75%. Only moderate grain-growth inhibition was observed [30].

3.2.5.2 Activated sintering and liquid phase sintering. The sintering rate of ceramics can be increased by additives often known as sintering aids. In activated sintering a very small amount of such an additive, usually $\leqslant 1\%$ is sufficient. The activator is thought to seggregate at particle surfaces and interparticle boundaries and activate the boundary diffusion processes. For example an effective additive for alumina is MgO [31] which forms a grain-boundary film and thereby enhances boundary diffusion as well as reducing the boundary mobility (i.e. grain growth) (Figure 3.5c). Other sintering aids that have been found effective are Al, C and B for SiC, and alumina and yttria for silicon nitride and zirconia.

For liquid phase sintering a larger addition of the sintering aid is made (3–30 vol%) [32]. This is chosen to have lower melting point than the main constituent and when in the molten state to wet it. The sintering temperature is set just above the melting point of the added phase so that during sintering it forms a liquid phase that wets the solid particles. Thus the pores in the compact are largely surrounded by the liquid phase and the driving force for sintering is the liquid surface energy. The first stage of sintering, namely rearrangement is very effective since the liquid fills pores, breaks up particle contacts and removes interparticle friction. With high liquid fractions, full density can be achieved almost entirely by rearrangement. When a solid skeleton develops before full densification, the diffusion stage of densification is often enhanced since it occurs by liquid phase diffusion.

Frequently (though not always), the enhanced diffusion also causes rapid grain growth. As in solid state sintering the grain growth can be effectively inhibited by introducing particles of a second solid phase [33, 34] which does not normally have a significant effect on the liquid phase densification mechanisms. However, again whiskers and short fibres appear to inhibit the densification [35].

Liquid phase sintering is an effective means of achieving practically porefree microstructures without the assistance of external pressure.

Important examples are the sintering of silicon nitride with a glassy liquid phase (a typical composition is $2\,wt\%$ alumina $+ 6\,wt\%$ yttria which form the glass phase together with some of the nitride) and the sintering of SiC with a silicon liquid phase. The main drawback of liquid phase sintering is that the useful application temperature of the material is limited by the presence of the low melting-point material. One way to circumvent this is to devise systems in which the liquid phase is transient, for example through dissolution in the solid phase as sintering progresses.

3.2.5.3 Reaction bonding. Originally, reaction bonding was devised as a method of preparing silicon nitride. A compact of pure silicon is sintered in a nitrogen atmosphere at 1300–1400 °C and thereby reacts to form a body of silicon nitride (sometimes termed RBSN) containing about 15–20% porosity [4]. Since the body takes up nitrogen it exhibits no shrinkage; the reaction product instead fills the porosity. The principle can be used to produce denser material without shrinkage by sintering a balanced mixture of silicon and silicon nitride in nitrogen [36]. A similar process has been reported for the reaction bonding of an aluminium/alumina mixture [37]. Reaction bonding of SiC is achieved by infiltrating, with molten silicon, compacts consisting of SiC mixed with free carbon. Again little shrinkage occurs. Reaction bonding methods have potential for producing fibre reinforced ceramics since by eliminating shrinkage the problem of shrinkage inhibition by fibres is avoided.

3.2.5.4 Reactive sintering and combustion synthesis. The term reactive sintering implies that two or more constituents in a compact react during sintering to form a new phase or phases. The reaction is normally exothermic and can contribute to an enhancement of sintering. In some cases the reaction is so exothermic that it can generate sufficient heat to cause self-sintering without external heating except that required for initiating the reaction. This is the basis of combustion synthesis which if properly controlled can produce a relatively dense compact of the synthesised reaction product [38]. Alternatively, pressure can be applied immediately after the reaction thus utilising the heat of reaction in further consolidation. Two examples of composites produced are:

- Reaction sintering [39]:

$$3TiO_2 + 4AlN \rightarrow 2Al_2O_3 + 2TiN + N_2 \qquad (3.2)$$

- Combustion synthesis [40]:

$$3TiO_2 + 4Al + 3C \rightarrow 3TiC + 2Al_2O_3 \qquad (3.3)$$

3.2.5.5 Pressure assisted sintering. As already indicated it is difficult to achieve full density in many ceramics and ceramic composites by pressureless sintering. This is particularly so for whisker and short-fibre reinforced composites. Greater success is achieved if the sintering is enhanced by simultaneous external pressure. Since sintering temperatures and times can be reduced, less grain growth occurs. The two main types of pressure assisted sintering, namely uniaxial hot pressing and hot isostatic pressing (HIP) are illustrated in Figures 3.6a, b.

Uniaxial hot pressing is essentially uniaxial pressing in a high temperature furnace [41]. Temperatures up to 2000 °C and pressures up to about 50 MPa are common. The die and punches are usually of graphite, sometimes coated with boron nitride, and pressing is carried out in an inert atmosphere to protect tooling and furnace parts. The pressed component is restricted to simple geometries, usually cylinders or discs with up to around 20 cm diameter. More complex components have to be machined. The method is relatively costly.

In the HIP process, pressures of up to 200 MPa are applied via a gas that surrounds the component in a gas-tight furnace chamber. Since the pressure is isostatic, large and complex components can be pressed. The component has to be surrounded by a gas-tight container. In the glass encapsulation technique this is prepared by coating the green compact (formed by one of the forming methods described above) with a glass powder slurry. Upon heating (in vacuum) the glass softens and forms a gas-tight skin [42]. A convenient alternative is provided by a sinter-HIP sequence in which the component is first pressureless sintered (in a vacuum) to give a closed porosity and then subjected to HIP without requiring encapsulation.

3.3   Whisker and short-fibre reinforced composites

In earlier sections it has become apparent that whisker and short-fibre composites are difficult to produce using conventional cold forming

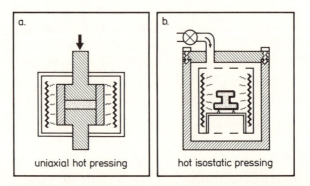

Figure 3.6   Schematic illustrations of (a) uniaxial hot press, (b) hot isostatic press.

techniques followed by pressureless sintering. The main source of difficulty is the poor packing characteristics of particles with high aspect ratio (see section 3.2 and Figure 3.3). However, these composites can be prepared using conventional processing with relatively straightforward modifications provided pressure-assisted sintering methods such as hot pressing are used for consolidation. It is this fact together with the availability and superior properties of SiC whiskers that has led to their rapid development (see chapter 9).

Most currently used processing routes for whisker reinforced ceramics include the following stages:

(1) whisker 'cleaning', i.e. wet dispersion followed by sedimentation [43] or sieving [44] to remove whisker agglomerates and other impurities;
(2) wet mixing of matrix powder and whiskers, typically by ball milling [43, 45], ultrasonic homogenising [45] or high speed homogenising [46]. Some success has also been achieved in stabilising whisker suspensions [47, 48];
(3) sintering (hot pressing); uniaxial hot pressing is the most common method used both in research and development work and in the commercial production of cutting tools.

Interest is being focused on other methods to produce more complex components with near net shape. This has led to a need to study alternative forming methods such as injection moulding and slip casting in association with for example hot isostatic pressing. Some reported examples are:

- cold isostatic + sintering to closed porosity + post HIP (without encapsulation), i.e. sinter/HIP [49];
- cold isostatic pressing or slip casting + HIP (glass encapsulation) [46, 50];
- injection moulding + HIP (glass encapsulation) [51];
- slip casting [52] + sinter/HIP [53];
- slip casting + reaction bonding [54].

The sinter/HIP approach has so far only been possible with low whisker contents (10–15 vol%) because at higher contents an interconnected, undeformable whisker skeleton is formed which hinders further shrinkage. To obtain maximum toughness and strength around 25–30 vol% is needed (see chapter 9). Furthermore, to accomplish the pressureless sintering stage, sintering aids are added which results in less favourable high temperature properties. With encapsulated HIP, the composites can be sintered to full density with relatively small amounts of sintering aids [55].

Producing complex components with for example slip casting [52] or injection moulding [51] introduces special problems associated with variations of whisker orientation in the component. Such variations can cause corresponding variations in shrinkage during sintering leading to distortion in the component, as illustrated in Figure 3.7.

A further problem associated with small-diameter whiskers is a potential health hazard akin to that experienced with asbestos. This necessitates the added expense of special handling facilities during processing stages prior to sintering and has diverted interest to the use of platelet reinforcement. However, the fully sintered materials are unlikely to be hazardous [56].

3.4 Long-fibre composites

It is convenient to distinguish between those composites prepared by impregnation of a continuous, multifilament yarn of fibres with matrix (most commonly in the form of a powder slurry) and those prepared by infiltration of a fibre preform of predetermined shape and usually with a multiaxial fibre geometry. In the former case the impregnated yarn can be laid up into various geometries prior to consolidation.

3.4.1 *Long-fibre/slurry, hot pressed composites*

A class of ceramic-matrix composites has been developed utilising the fact that a mixture of glass or ceramic powder and fibres can be densified to full density at a temperature low enough not to damage the fibres in cases where the matrix powder has a relatively low melting point and densifies by viscous flow when pressure is applied during sintering [57–60]. Composite laminate structures, for example, can be produced as illustrated in Figure 3.8. A fibre yarn is passed through a slurry containing matrix powder and additives (binder, dispersant etc.) and is then wound onto a spool where it can dry. Unidirectional 'pre-pregs' can then be cut, stacked and hot pressed after burn-out of the binder. Although a two-dimensional lay-up is preferred for hot pressing, the slurry technique can be used for other configurations such as filament wound structures.

Figure 3.7 Source of shrinkage distortion during sintering of injection moulded, whisker reinforced composite.

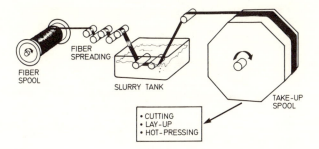

Figure 3.8 Schematic illustration of slurry method for preparing long-fibre reinforced ceramics.

Hot pressing is carried out at temperatures where viscous flow is the dominant densification mechanism. The lack of fibres resistant to high-temperature degradation or oxidation has led to relatively low melting point materials, such as glasses and glass ceramics, being the most successful matrices. The latter can be melted and formed as glass but crystallise during appropriate heat treatment to yield a polycrystalline ceramic [58, 59]. A wide range of melting points and thermal expansions can be obtained by adjustment of the matrix composition. The most commonly studied matrices are taken from the lithium aluminium silicates (LAS) and magnesium aluminium silicates (MAS) [58]. It is also possible to use the slurry/hot press method with matrices that densify by liquid-phase sintering. For example, silicon nitride matrix composites (with high content of sintering aids) reinforced with C-fibres [62] or SiC monofilaments (CVD) [60, 61] have been hot-pressed to full density. Needless to say the high temperature properties of such composites will be limited by the low melting points of the matrices.

3.4.2 Long-fibre composites produced by infiltration of preforms

3.4.2.1 The preform principle. Using the principle of infiltration of porous bodies with vapours, liquids or suspensions it has been found possible to produce composites on the basis of multidirectional and woven preforms, both two- and three-dimensional. Two- and three-dimensional preforms can be prepared from long fibres using textile production techniques such as weaving, stitching, knitting and braiding (see Figure 3.9). These have been applied successfully on such fibres as multifilament C, SiC and Al_2O_3. Since such preforms are often flexible they can be formed into larger structures. A potential drawback of composites produced from such preforms is the existence of large local variations in volume fraction and relatively large regions of fibre-free matrix. Nevertheless, overall fibre volume fractions of around 50% can

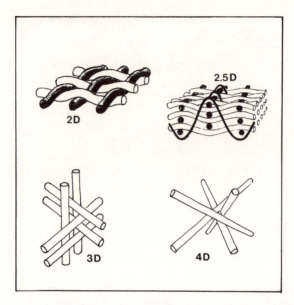

Figure 3.9 Examples of multidirectional woven fibre structures that can be used as a basis for preforms (the term D refers here to direction rather than dimension).

be attained. A thorough review of preform preparation and architecture has been presented by Ko [63]. The principles of some of the more important infiltration techniques are outlined below.

3.4.2.2 Chemical vapour infiltration (CVI). Chemical vapour deposition (CVD) which is a well-established technique for depositing thin ceramic coatings [15] can be modified and used to infiltrate a porous preform and deposit a ceramic matrix on fibres (see Figure 3.10). The ceramic is formed *in situ* by the chemical reaction in a gas mixture passing through the preform heated to the appropriate reaction temperature, usually somewhat above 1000 °C. By choice of suitable reactant gases, almost any ceramic compound can be deposited, the most common reactions involving the reduction of a chloride with hydrogen. Some examples are given in Table 3.1, a fuller list of reactions is provided in [15] and specific composites are described in chapter 8. Perhaps the most important example is the preparation of SiC/SiC composites by deposition of SiC using a reaction in which methyltrichlorosilane decomposes in the presence of hydrogen at around 1000 °C [64]:

$$CH_3SiCl_3 + H_2 \longrightarrow SiC + 3HCl + H_2 \tag{3.4}$$

The gas temperatures, pressures and flow characteristics as well as the substrate temperature have to be optimised in such a way as to delay

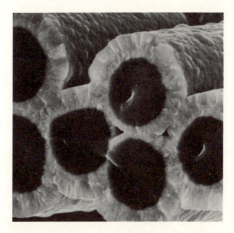

Figure 3.10   Example of chemical vapour infiltration, TiC deposited on C fibres.

closing the porosity at the preform surface before a sufficiently high density is attained in the interior. At the same time the effect of these parameters on the structure (and therefore properties) of the deposited ceramic must be taken into consideration. The most favourable microstructure for good mechanical properties consists of very fine equiaxed grains. The deposition of such a structure is favoured by low deposition temperature and high reactant supersaturation giving low diffusion rates and dense nucleation formation. Low temperatures also minimise the risk of fibre degradation. An additional feature of the CVI technique is that multiple layers can be deposited permitting the creation of interfacial layers.

A typical process for the production of a component by CVI is illustrated schematically in Figure 3.11. The infiltration can be carried out as a batch process in which many components are loaded into a large reaction chamber. The infiltration time, with intermediate surface removal is typically a matter of days or weeks. Small components have,

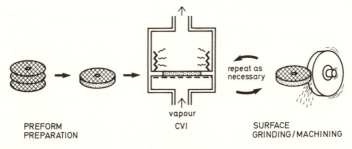

Figure 3.11   Schematic illustration of CVI preparation of a fibre reinforced ceramic component.

however, been produced in a forced flow set-up in less than 24 h [65]. Densities in the range 80–90% of theoretical density with < 10% open porosity are normally attained with the CVI method.

3.4.2.3 Polymer infiltration and pyrolysis. Fibre preforms can be infiltrated with liquid polymers, either molten or in solution, which are then pyrolised to leave a ceramic deposit. For example, pyrolysis of polycarbosilanes [66, 67] and polysilazanes [68–70] can be utilised to form matrices of SiC and Si_3N_4, respectively. The method has been used extensively in the production of carbon/carbon composites [71] and has more recently been extended to B-N-containing polymers [72], in particular for the deposition of oxidation-resistant, low bond-strength BN coatings on fibres [73]. The advantage of pyrolysis methods is the low process temperature, typically $\leqslant 1000\,°C$. Moreover, it is easier to infiltrate a fibre preform with a homogenous liquid than with a powder slurry. A drawback is again long processing times; pyrolysis yields are typically 70–85% retained weight and repeated impregnation/pyrolysis steps are needed to achieve adequate densities. Another drawback is the high cost of polymer precursors. Carbosilanes or silazanes are today produced only in small cost-ineffective quantities. It is however possible to synthesise these precursors using cheap waste products from the silicone industry which would reduce the cost significantly [68]. Precursors for carbon (phenolic resins, petroleum pitch) are already fairly low-cost chemicals.

3.4.2.4 Sol-gel infiltration. Sols have been used successfully to infiltrate preforms. Following infiltration they are gelled by drying or by adjusting the temperature [74]. A low density gel is formed which after drying can be sintered at considerably lower temperatures than ceramic powder compacts. As is the case for polymer pyrolysis, drawbacks are high cost and the need for repeated impregnation/sinter steps. The sol-gel technique has so far been used mainly to produce oxide matrices such as alumina, zirconia, silica and mullite [9, 74, 75].

3.4.2.5 Combined methods. Both polymer pyrolysis and sol-gel techniques can be combined with CVI or powder methods. For example, a first infiltration can be performed with a polymer or sol carrying a dispersed ceramic powder while subsequent infiltrations could be with pure polymer or sol. The final infiltration could be CVI filling the smallest pores and providing an oxidation resistant, high-purity surface coating [76].

3.4.2.6 Reaction bonding. The preform may also be infiltrated with a slurry or liquid which after deposition is reacted with a gas or liquid to form the desired ceramic matrix by reaction bonding [77]. Preforms have, for example, been infiltrated with silicon powder slurries and, after drying, sintered in nitrogen to form reaction bonded silicon nitride. Composites containing monofilament, CVD SiC fibres have been successfully prepared in this way [78]. Multifilament, polymer-precursor SiC fibres such as Nicalon (see chapter 2) which contain excess carbon are more reactive towards the silicon at the usual nitridation temperatures and degrade seriously [79]. However, by decreasing the nitridation time and temperature or applying a protective coating on the fibres it is possible to obtain such composites with relatively undamaged fibres [80, 81]. An important advantage of reaction bonding is that such reactions usually involve volume increases, thus more effectively filling the preform pores than, for example the polymer-pyrolysis and sol-gel techniques.

3.4.2.7 Melt infiltration.   A new technique now being pursued is the infiltration of preforms with a melt of the matrix material (a process analogous to the squeeze casting of metal-matrix composites) [58, 82]. This is a promising technique since high-density, flaw-free composites can be formed in a single processing step, with small dimensional change from preform to final product and with a wide choice of preform geometry. However, the scope of the process will be limited by fibre-damaging chemical interactions between the fibre and the melt at the high temperatures required and by the high viscosities of molten ceramics and glasses. The feasibility of the method has been demonstrated using a SiC whisker preform and low melting-point glasses [82].

3.4.2.8 Directed melt oxidation/nitridation (DIMOX^{TM}).   A novel method of preform infiltration involves the so-called directed oxidation (or nitridation) of molten metals [83]. The most well-documented example is that of aluminium. It has been found that in the presence of certain minor alloying additions, for example Mg, alumina can grow very rapidly from the surface of a melt. The process involves the maintenance of metal channels in the oxide. It has further been found that if a fibre preform is placed on top of the melt then the oxide will grow up through the preform to form a dense composite without disturbing the fibre geometry [84]. A consequence of the process is that the matrix will always contain a residue of metallic phase.

References

1. Reed, J. S. (1988) *Introduction to the Principles of Ceramic Processing*, Wiley, New York.
2. Rhodes, W. H. and Natansohn, S. (1989) Powders for advanced structural ceramics, *Am. Ceram. Soc. Bull.* 68, 1804–1812.

3. Srinivasan, M. (1989) The silicon carbide family of structural ceramics, *Treatise Mater. Sci. Technol.* 29, 99–159.
4. Torti, L. T. (1989) The silicon nitride and sialon families of structural ceramics, *Treatise Mater. Sci. Technol.* 29, 161–194.
5. Sarin, V. K. (1988) On the α- to β-phase transformation in silicon nitride, *Mater. Sci. Eng.* A105/106, 151–159.
6. Real, M. W. (1986) Freeze drying alumina powders, *Proc. Br. Ceram. Soc.* 38, 59–66.
7. Zhang, S. C., Messing, G. L. and Borden, M. (1990) Synthesis of solid, spherical zirconia particles by spray pyrolysis, *J. Am. Ceram. Soc.* 73, 61–67.
8. Slamovich, E. B. and Lange, F. F. (1988) Spherical zirconia particles via electrostatic atomization: fabrication and sintering characteristics, *Mater. Res. Soc. Symp. Proc.* 121, 257–262.
9. Colomban, P. (1989) Gel technology in ceramics, glass ceramics and ceramic-ceramic composites, *Ceram. Int.* 15, 23–50.
10. Segal, D. (1989) *Chemical Synthesis of Advanced Ceramics*, Cambridge University Press, New York.
11. Jones, R. W. (1989) *Fundamental Principles of Sol-Gel Technology*, Institute of Metals, London.
12. Brinker, C. J. and Sherer, G. W. (1990). *Sol-Gel Science; The Physics and Chemistry of Sol-Gel Processing*, Academic Press Inc., Boston.
13. Montanaro, L. and Guilhot, B. (1989) Preparation of microspheres from an alumina-zirconia sol, *Am. Ceram. Soc. Bull.* 68, 1017–1020.
14. Hatakeyama, F. and Kanzaki, S. (1990) Synthesis of monodispersed spherical β-silicon carbide powder by a sol-gel process, *J. Am. Ceram. Soc.* 73, 2107–2110.
15. Stinton, D. P., Besmann, T. M. and Lowden R. A. (1988) Advanced ceramics by chemical vapor deposition techniques, *Am. Ceram. Soc. Bull.* 67, 350–355.
16. Kato, A. (1987) Vapor-phase synthesis of ceramic powders, *Advances in Ceramics.* 21, 181–192.
17. Anderson, H., Kodas, T. T. and Smith, D. M. (1989) Vapor-phase processing of powders: plasma synthesis and aerosol decomposition, *Am. Ceram. Soc. Bull.* 68, 996–1000.
18. Bostedt, E., Persson, M. and Carlsson, R. (1989) Colloidal processing through surface modification, in: *Euro-Ceramics, Processing of Ceramics*, eds. de With, G., Terpstra, R. A. and Metselaar, R., Elsevier Applied Science, London, pp. 1.140–1.144.
19. Carlström, E. and Lange, F. F. (1984) Mixing of flocced suspensions, *J. Am. Ceram. Soc.* 67, C-169–170.
20. Sanders, W. A., Kiser, J. D. and Freedman, M. R. (1989) Slurry-pressing consolidation of silicon nitride, *Am. Ceram. Soc. Bull.* 68, 1836–1841.
21. Cesarano, J., Aksay, I. A. and Bleier, A. (1988) Stability of aqueous alpha alumina suspensions with poly(methacrylic acid) polyelectrolyte, *J. Am. Ceram. Soc.* 71, 250–255.
22. Ruys, A. J. and Sorrell, C. C. (1990) Slip casting of high purity alumina using sodium carboxymethylcellulose as deflocculant/binder, *Am. Ceram. Soc. Bull.* 69, 828–832.
23. Persson, M., Forsgren, A., Carlström, E., Käll, L., Kronberg, B., Pompe, R. and Carlsson R. (1987) Steric stabilization of silicon carbide slips, in: *High Tech Ceramics*, ed. Vincenzini, P., Elsevier Science Publishers, Amsterdam, pp. 623–632.
24. Lehmann, J., Müller, B. and Ziegler, G. (1989) in: *Euro-Ceramics, Processing of Ceramics*, eds. de With, G., Terpstra, R. A. and Metselaar, R., Elsevier Applied Science, London, pp. 1.196–1.200.
25. German, R. M. (1990) *Powder Injection Moulding*, MPI, New Jersey.
26. Milewski, J. V. (1986) Efficient use of whiskers in the reinforcement of ceramics, *Adv. Ceram. Mater.* 1, 36–41.
27. Exner, H. E. (1979) Principles of single phase sintering, *Rev. Powder Met. Phys. Ceram.* 1, 1–251.
28. Kellet, B. J. and Lange, F. F. (1989) Thermodynamics of densification: 1, sintering of simple particle arrays, equilibrium configurations, pore stability and shrinkage, *J. Am. Ceram. Soc.* 72, 725–734.

29. French, J. D., Harmer, M. P., Chan, H. M. and Miller, G. A. (1990) Coarsening resistant dual phase interpenetrating microstructures, *J. Am. Ceram. Soc.* 73, 2508–2510.
30. Lee, H. W. and Sacks, M. D. (1990) Pressureless sintering of SiC whisker reinforced alumina composites 1, effect of matrix powder surface area, *J. Am. Ceram. Soc.* 73, 1884–1893.
31. Bennison, S. J. and Harmer, M. P. (1990) Effect of magnesia solute on surface diffusion in sapphire and the role of magnesia in the sintering of alumina, *J. Am. Ceram. Soc.* 73, 833–837.
32. German, R. M. (1985) *Liquid Phase Sintering*, Plenum Press, New York.
33. Stephenson, I. M. and White, J. (1967) Factors controlling microstructures and grain growth in two-phase and three-phase systems, *Trans. Brit. Ceram. Soc.* 66, 443–483.
34. Warren, R. (1972) Effects of the carbide composition on the microstructure of cemented binary carbides, *Plansceberichte Pulvermet.* 20, 299–317.
35. Tiegs, T. N. and Dillard, D. M. (1990) Effect of aspect ratio and liquid phase content on the densification of alumina-silicon carbide whisker composites, *J. Am. Ceram. Soc.* 73, 1440–1442.
36. Pompe, R., Hermansson, L. and Carlsson, R. (1982) Fabrication of low-shrinkage silicon nitride material by pressureless sintering, in: *Engineering with Ceramics II*, ed. Davidge, R. W., British Ceram. Soc., Stoke on Trent, pp. 65–74.
37. Claussen, N., Le, T. and Wu, S. (1989) Low-shrinkage reaction bonded alumina, *J. Am. Ceram. Soc.* 5, 29–35.
38. Munir, Z. A. (1988) Synthesis of high temperature materials by self-propagating combustion methods, *Am. Ceram. Soc. Bull.* 67, 342–349.
39. Mukerji, J. and Biswas, S. K. (1990) Synthesis, properties and oxidation of alumina-titanium nitride composites, *J. Am. Ceram. Soc.* 73, 142–145.
40. Rabin, H. R., Korth, G. E. and Williamson, R. L. (1990) Fabrication of titanium carbide-alumina composites by combustion synthesis and subsequent dynamic consolidation, *J. Am. Ceram. Soc.* 73, 2156–2157.
41. Richerson, D.W. (1982) *Modern Ceramic Engineering*, Marcel Dekker, New York.
42. Larker, H. T. (1983) in: *Progress in Nitrogen Ceramics*, ed. Riley, F. L., Nijhoff, Boston, pp. 717–724.
43. Lundberg, R., Nyberg, B., Williander, K., Persson, M. and Carlsson, R. (1987) Processing of whisker reinforced ceramics, *Composites* 18, 125–127.
44. Tamari, N., Ogura, T., Kinoshita, M. and Toibana, Y. (1982) Fabrication of SiC whisker/silicon nitride composite materials and their physical properties, *GIRIO Bull.* 33, 129–134.
45. Tiegs, T. N. and Becher, P.F. (1986) Whisker reinforced ceramic composites, in: *Ceramic Materials and Components for Engines*, eds. Bunk, W. and Hausner, H., DKG, Bad Honnef, Germany, pp. 193–200.
46. Lundberg, R., Kahlman, L., Warren, R., Pompe, R. and Carlsson, R. (1987) SiC whisker reinforced Si_3N_4 composites, *Am. Ceram. Soc. Bull.* 66, 330–333.
47. Konszlowicz, K. J. (1989) Dispersion of SiC whiskers in water, in: *Euro-Ceramics Processing of Ceramics*, eds. de With, G., Terpstra, R. A. and Metselaar, R., Elsevier Applied Science, London, pp. 1.169–1.173.
48. Mutsuddy, B. C. (1990) Electrokinetic behavior of aqueous silicon carbide whisker suspensions, *J. Am. Ceram. Soc.* 73, 2747–2749.
49. Tiegs, T. N. and Becher, P. F. (1987) Sintered alumina-SiC whisker composites, *Am. Ceram. Soc. Bull.* 66, 339–342.
50. Lundberg, R., Nyberg, B., Williander, K., Persson, M. and Carlsson, R. (1988) in: *Proc. 1st Int. Conf. HIP, 1987*, ed. Garvare, T., Centek Publications, Luleå, Sweden, pp. 323–327.
51. Buljan, S. T., Baldoni, J. G., Niel, J. and Zilberstein, G. (1987) Dispersoid toughened silicon nitride composites, ORNI report/Sub. 85-22011/1.
52. Hoffman, J. Nagel, A., Greil, P. and Petzow, G. (1989) Slip casting of SiC whisker reinforced silicon nitride, *J. Am. Ceram. Soc.* 72, 765–769.
53. Hoffman, M. J., Greil, P. and Petzow, G. (1989) Pressureless sintering of SiC-whisker reinforced silicon nitride, in: *Science of Ceramics 14*, ed. Taylor, D., Institute of Ceramics, pp. 825–830.

54. Starr, T. L. and Harris, J. N. (1986) Fabrication of slip cast, reaction sintered silicon nitride composites, in: *Ceramic Materials and Components for Engines*, eds. Bunk, W. and Hausner, H., DKG, Bad Honnef, Germany, pp. 217–224.

55. Takemura, H., Miyamoto, Y. and Koizumi, M. (1988) Fabrication of dense Si_3N_4-SiC whisker composite without additives by HIP, in: *Proc. 1st Int. Conf. on HIP, 1987*, ed. Garvare, T., Centek Publications, Luleå, Sweden, pp. 329–335.

56. Bogoroch, R. and Luck, S. R. (1988) Workplace handling requirements and procedures for ACMC silicon carbide whiskers based on a subchronic inhalation study in rats, in: *Whisker and Fiber-Toughened Ceramics*, eds. Bradley, R. A., Clarke, D. E., Larsen, D. C. and Stiegler, J. O., ASM International, pp. 81–89.

57. Sambell, R. A. J., Bowen, D. H. and Phillips, D. C. (1972) Carbon fibre composites with ceramic and glass matrices, *J. Mater Sci.* 7, 663–675.

58. Prewo, K. M., Brennan, J. J. and Layden, G. K. (1986) Fiber reinforced glasses and glass ceramics for high performance applications, *Am. Ceram. Soc. Bull.* 65, 305–313 and 322.

59. Brennan, J. J. and Prewo, K. M. (1982) Silicon carbide fibre reinforced glass ceramic matrix composites exhibiting high strength and toughness, *J. Mater. Sci.* 17, 2371–2383.

60. Shetty, D. K., Pascucci, M. R., Mutsuddy, B. C. and Wills, R. R. (1985) SiC monofilament reinforced silicon nitride matrix composites, *Ceram. Eng. Sci. Proc.* 6(7-8), 632–645.

61. Swab, J. J., Quinn, G. D. and Snona, P. J. (1990) Mechanical behaviour of a SiC-fibre/Si_3N_4 composite, US Army Report, MTL TN 90-2.

62. Lundberg, R., Pompe, R. and Carlsson, R. (1986) HIPed carbon fibre reinforced silicon nitride composites, *Ceram. Eng. Sci. Proc.* 9(7-8), 901–906.

63. Ko, F. K. (1989) Preform fiber architecture for ceramic matrix composites, *Am. Ceram. Soc. Bull.* 68, 401–414.

64. Besmann, T. M., Sheldon, B. W. and Kaster, M. D. (1990) Temperature and concentration dependence of SiC deposition on Nicalon fibres, *Surf. Coatings Technol.* 43/44, 167–175.

65. Caputo, A. J., Stinton, D. P., Lowden, R. A. and Besmann, T. M. (1987) Fiber-reinforced SiC composites with improved mechanical properties, *Am. Ceram. Soc. Bull.* 66, 368–372.

66. Fitzer, E. and Gadow, R. (1986) Fibre-reinforced silicon carbide, *Am. Ceram. Soc. Bull.* 65, 326–335.

67. Boisvert, R. P. and Diefendorf, R. J. (1988) Polymeric precursor SiC matrix composites, *Ceram. Eng. Sci. Proc.* 9(7-8), 873–880.

68. Seyferth, D. and Wiseman, G. H. (1986) A novel polymeric organisilazane precursor to Si_3N_4/SiC ceramics, in: *Science of Ceramic Chemical Processing*, eds. Hench, L. L. and Ulrich, E. R., John Wiley and Sons, New York, pp. 354–362.

69. Sirieix, F., Goursat, P., Lecomte, A. and Dauger, A. (1990) Pyrolysis of polysilazanes: relationship between precursor architecture and ceramic microstructure, *Composites Sci. Technol.* 37, 7–19.

70. Lundberg, R. and Goursat, P. (1989) Silicon carbonitride ceramic matrix composites by polymer pyrolysis, *Proc. 3rd European Conf. Composite Materials*, EACM, Bordeaux, France, pp. 93–98.

71. Buckley, J. D. (1988) Carbon-carbon, an overview, *Am. Ceram. Soc. Bull.* 67, 364–368.

72. Lightfoot, A., Rees, W. R. and Haggerty, J. S. (1988) Boron-containing ceramic materials derived from polymeric precursors: material characteristics, *Ceram. Eng. Sci. Proc.* 9(7-8), 1012–1030.

73. Paciorek, K. J. L. and Kratzer, R. H. (1988) Boron nitride preceramic polymer studies, *Ceram. Eng. Sci. Proc.* 9(7-8), 993–1000.

74. Zelinski, B. J. and Uhlman, D. R. (1984) Gel technology in ceramics, *J. Phys. Chem. Solids* 45, 1069–1090.

75. Pierre, A. C., Uhlman, D. R. and Hordonneau, A. (1986) Ceramic composites made by sol-gel processing, *Rev. Int. Hautes Temp. Refract.* 23, 29–35.

76. Strife, J. J. and Sheehan, J. A. (1988) Ceramic coatings for carbon-carbon composites, *Am. Ceram. Soc. Bull.* 67, 369–374.

77. Washburn, M. E. and Coblenz, W. S. (1988) Reaction formed ceramics, *Am. Ceram. Soc. Bull.* 67, 356–363.
78. Corbin, N. D., Rossetti, G. A. and Hartline, S. D. (1986) Microstructure/property relationship for SiC filament reinforced RBSN, *Ceram. Eng. Sci. Proc.* 7(7–8), 958–968.
79. Fischbach, D. B. and MacLaren, D. (1982) Exploratory research on silicon nitride composites, NASA-Report, DOE ET 13389-T1.
80. Corbin, N. D., Wilkens, C. A. and Hartline, S. D. (1986) The influence of fiber coatings on RBSN matrix composites, presented at DOD/NASA sponsored Ceramic Matrix Composites meeting (restricted session), Cocoa Beach, USA.
81. Lundberg, R., Pompe, R. and Carlsson, R. (1987) Silicon carbide fibre reinforced nitrided silicon nitride composites, *Proc. ICCM VI,* Vol. 2, Elsevier, London, pp. 33–39.
82. Hillig, W. B. (1988) Properties and structure of melt infiltrated composites, *Ceram. Eng. Sci. Proc.* 9(7-8), 755–758.
83. Newkirk, M. S., Urquhart, A. W., Zwicker, H. R. and Breval, E. (1986) Formation of Lanxide™ ceramic composite materials, *J. Mater. Res.* 1, 81–89.
84. Barron-Antolin, P., Schiroky, G. H. and Andersson, C. A. (1988) Properties of fibre-reinforced alumina matrix composites, *Ceram. Eng. Sci. Proc.* 9(7-8), 759–766.

4 Fundamental aspects of the properties of ceramic-matrix composites

R. WARREN

4.1 Introduction

This chapter provides a brief survey of the principles that determine the mechanical and physical properties of ceramic-matrix composites. As already discussed briefly in chapter 1, the main factors affecting the properties are the constituent phase properties, the reinforcement morphology and the reinforcement/matrix interface and this should become clearer in this chapter. The properties of most concern for structural ceramics are the stress–strain and fracture behaviour particularly with respect to fracture toughness and methods of improving toughness and reliability.

These subjects have received most attention in the literature and are dealt with in some detail in the first part of the chapter. Following this, other properties of practical importance such as high-temperature deformation and failure, fatigue, wear and oxidation are discussed albeit somewhat more briefly. Finally, simple predictive laws for selected relevant physical properties are presented.

The approach of the chapter is as general as possible. The reader is advised to consult later chapters to find examples of specific systems that illustrate the general presentation. Similarly, the chapter does not attempt to provide an exhaustive reference list. Instead reference is made to useful reviews or where appropriate, recent theoretical advances.

4.2 Time-dependent stress–strain and fracture behaviour

4.2.1 *Fracture mechanisms and basic fracture mechanics*

It is helpful for the subsequent discussion to distinguish three main types of fracture behaviour that can be observed in ceramic composites (Figure 4.1). The first of these is a classical linear elastic brittle fracture with rapid, uncontrolled crack growth. The slope of the stress–strain curve is determined by the elastic stiffness of the sample while the fracture stress is determined in the simplest case by the fracture

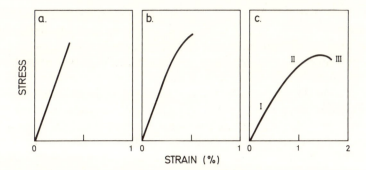

STRAIN (%)

Figure 4.1 Three types of stress–strain curve observed in ceramic composites. Typical strain levels are indicated.

toughness of the material and the largest effective defect in the sample:

$$\sigma_F = Y.K_c/c^{1/2} \tag{4.1}$$

where, here, the fracture toughness is expressed in terms of the critical stress intensity factor, c is the defect size (linear) and Y is a dimensionless factor describing the sample and stress distribution geometry.

The second type of fracture (Figure 4.1b) is also brittle but final failure is preceded by some sub-critical crack growth evidenced by a fall in stiffness. Such behaviour can occur simply because the crack advances into a falling stress field (rising Y) but in composites it frequently occurs because the nature of the crack changes as the crack extends leading to a reduction in the stress intensity at the crack tip and therefore an apparent rise in the fracture toughness (the phenomena is commonly termed R-curve behaviour, R being the toughness expressed as the fracture surface energy). Possible mechanisms leading to such toughening in composites include crack deflection, crack bridging, fibre pull-out, microcracking and phase transformation. These are described in more detail in later sections of this chapter. It should be noted that the existence of R-curve behaviour makes the definition and measurement of a unique fracture toughness value problematical as will become clearer later (section 4.2.9) [1].

The sub-critical crack growth described here should not be confused with time-dependent stable crack growth caused by reaction of the environment with the crack tip under static fatigue conditions (section 4.3.1) or with stable crack growth associated with plastic or creep deformation at the crack tip at elevated temperatures (section 4.3.2).

The third type of fracture (Figure 4.1c) is non-brittle in character and is observed in axially loaded, uniaxial long-fibre composites having a weak fibre/matrix bond. The stress–strain curve is linear (region I) up to a strain corresponding to multiple cracking in the matrix. The fibres survive this matrix failure and continue to sustain the composite load. The ultimate failure of the composite occurs by the non-simultaneous failure (region II) and subsequent pull-out of the fibres (region III). This behaviour is desirable since it renders the material insensitive to defects and catastrophic failure is avoided. However the matrix cracking is irreversible damage which may be unacceptable in many applications. With a strong fibre/matrix bond, this pseudo-plasticity is lost; the fibres generally provide toughening but the fracture behaviour is brittle. Woven and laminated multidirectional long-fibre composites usually behave in a manner similar to the unidirectional composites. Off-axis laminates or fibre bundles fail at low strains while axial bundles fail in a gradual manner.

Before describing the various composite toughening processes in more detail it is appropriate to take up a number of topics of general relevance to fracture behaviour.

4.2.2 Statistical description of fracture reliability

The fracture strength of brittle materials exhibits considerable scatter between specimens of apparently identical geometry and loading conditions. This is largely due to a stochastic distribution of defect size and location in the material (i.e. c and Y are variable in equation (4.1)). Another important consequence of this is that the mean strength is dependent on sample size since the probability of large defects increases with sample volume. These effects apply to ceramic fibres as well as to bulk samples. In components, this scatter implies poor reliability to the designer.

The frequency distribution of strength values obtained in a representative number of tests in a material can usually be described mathematically by relatively simple expressions. The most commonly used is the so-called two-parameter Weibull distribution [2, 3]

$$P_s = \exp\left[- V(\sigma/\sigma_0)^m/V_0\right] \qquad (4.2)$$

where P_s is the probability that a given specimen subject to a uniform tensile load will survive beyond an applied stress σ. V is the specimen volume, V_0 is a reference volume which can conveniently be set to the unit of volume measurement (for fibres with uniform diameter V and V_0 can be replaced by lengths L and L_0), m and σ_0 are constants for the material. Roughly speaking σ_0 describes the stress level of the distribution while m is an inverse measure of the width of the distribution (i.e. the scatter).

Expression (4.2) is the cumulative form of a frequency distribution having a single maximum with the lower tail fixed at $\sigma = 0$ and the upper tail skewed towards high stress values. A plot of the expression in a diagram of $\ln\ln 1/P_s$ versus $\ln\sigma$ yields a straight line with slope m (a so-called Weibull plot). An example of such a plot for SiC fibres is shown in Figure 4.2 [4]. Experimental data frequently deviate from the simple Weibull distribution and better fit requires expressions with a larger number of adjustable parameters. Nevertheless, the expression has won wide acceptance; σ_0 and m values are extracted on a best-fit basis.

The mean strength is normally quite close to the median strength which is given by the value of σ at $P_s = 0.5$. Thus

$$\ln\bar{\sigma} \approx \ln\sigma_{\text{median}} = -(1/m)\ln(V/V_0) - 0.367/m + \ln\sigma_0 \qquad (4.3)$$

which yields the effect of specimen volume on mean strength. Values of the Weibull modulus, m, for ceramics and ceramic fibres lie typically in the range 5–25.

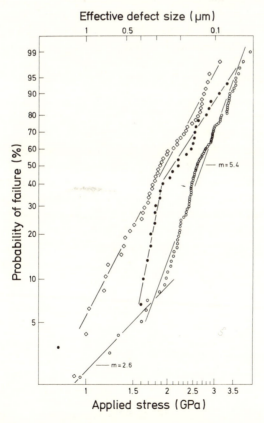

Figure 4.2 Weibull plot for three different samples of SiC (Nicalon) fibres [4]. (The vertical scale is linear with respect to $\ln\ln(1/P_s)$; it is converted to probability of failure for convenience of interpretation.)

In most practical situations, for example a bend test specimen or a loaded component, the sample will be subjected to a non-uniform stress distribution and an integrated form of equation (4.2) has to be applied. The expected influence of composite toughening mechanisms on the Weibull parameters is discussed in section 4.2.9.

4.2.3 Micromechanical description of stress–strain behaviour

Micromechanical models of composites are largely concerned with local stresses and strains in the microstructure during loading. As such they can provide first order estimates of fracture behaviour and provide a necessary background for more detailed analysis of fracture mechanisms.

4.2.3.1 Long, unidirectional fibres. It is convenient to begin by considering the axial loading of a unidirectional, long-fibre composite. This loading situation is particularly simple since matrix, fibre and composite strains are equal and consequently the stress–strain behaviour of the composite can be easily derived by reference to the individual stress –strain curves of the fibre and matrix as shown in Figure 4.3a. At any composite strain, ε_c, prior to fracture, the stresses in the fibre (σ_f) and matrix (σ_m) can be obtained from these curves or from

$$\sigma_f = E_f \varepsilon_c \tag{4.4a}$$

$$\sigma_m = E_m \varepsilon_c \tag{4.4b}$$

where E_f and E_m are the Young's moduli of fibre and matrix. It is to be noted that the figure applies to the majority of ceramic composites, namely the matrix has lower stiffness and fracture strength than the reinforcement and consequently is the first to reach its fracture strain. Up to this fracture strain the composite stress is given by

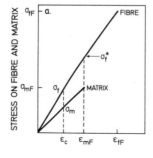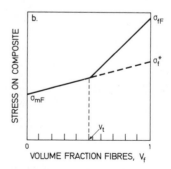

Figure 4.3 (a) Stress–strain relationships for brittle fibre and matrix. At composite strain, ε_c, the fibre and matrix stresses are σ_f and σ_m, respectively. (b) Composite fracture strength as a function of volume fraction for the matrix/fibre combination in (a).

$$\sigma_c = \sigma_f V_f + \sigma_m V_m \tag{4.5}$$

where V_f and V_m are the volume fractions of the constituents.

The axial Young's modulus of the composite is obtained by combining equations (4.4) and (4.5):

$$E_c = E_m V_m + E_f V_f \tag{4.6}$$

When the composite is loaded to the fracture strain of the matrix, the stresses in the two phases just prior to matrix fracture are $\sigma_{mF} = E_m \varepsilon_{mF}$ (where the subscript F refers to fracture) and the composite stress at this point is

$$\sigma_c \text{ (matrix fracture)} = \sigma_{mF} V_m + \sigma_f^* V_f \tag{4.7}$$

where σ_f^* is the stress in the fibre at $\varepsilon_f = \varepsilon_{mF}$. At sufficiently low fibre fractions, when the matrix fails, the fibres will not sustain the transfer of load and composite failure will occur simultaneously. The composite fracture stress will then be given by equation (4.7). This is indicated by the line to the left of V_t in Figure 4.3b. Above the transition volume fraction, $V_f > V_t$, the fibres will continue to carry load after matrix failure and final composite failure will be given by

$$\sigma_{cF} = \sigma_{fF} V_f \tag{4.8}$$

When the matrix fails it shrinks back from the fracture surface. Correspondingly, the matrix stress will relax close to the fracture surface but will rise to its full value, σ_{mF}, at a distance behind the fracture surface corresponding to a load transfer length that is inversely dependent on the fibre/matrix bond. Thus for constant or increasing composite load, fracture will continue to occur at a regular spacing throughout the entire length of the composite, a phenomenon known as multiple fracture. For simple frictional bonding, giving a constant interfacial shear stress, τ, the relaxation length and therefore the approximate multiple crack spacing is given by

$$l_r = \sigma_{mF} r V_m / 2\tau V_f \tag{4.9}$$

where r is the fibre radius.

The above simple rule of mixtures (ROM) model predicts the initial, multiple matrix cracking that is observed to occur prior to final failure in some long-fibre composites. According to the model this occurs at a composite stress given by the dotted line in Figure 4.3 and this is followed by sudden final fracture at the stress given by equation (4.8). The model however overlooks three important factors, namely the effect of the presence of the fibres on the fracture strain of the matrix, the effect of the matrix crack on the fracture of the fibres and the effect of the scatter in fibre strength.

In reality, matrix cracking will be inhibited because the fibres limit the extent of matrix strain relief that provides the driving energy for fracture. The extent of this effect is dependent on the strength of the fibre/matrix bond. For a frictional interface strength, τ, Aveston, Cooper and Kelly (ACK) [5] give for the matrix failure strain,

$$\varepsilon_{mF} = [6\tau\Gamma E_f V_f^2/E_m^2 E_c r V_m]^{1/3} \qquad (4.10)$$

where Γ is the specific fracture surface energy of the matrix. The matrix fracture stress and corresponding composite stress are given simply by $E_m\varepsilon_{mF}$ and $E_c\varepsilon_{mF}$, respectively.

Budiansky *et al*. [6] have extended this analysis to encompass a rigid fibre/matrix bond where the matrix shrinks back between the fibres by elastic relaxation only. Both analyses predict that above a certain volume fraction of fibres, the effective matrix strength and fracture strain are greater than the unreinforced values and that they increase with volume fraction. The ACK model predicts that the effect increases with fibre fraction. This leads to the prediction of composite behaviour given in Figure 4.4.

Up to a volume fraction given by point A (depending on the parameters in equation (4.10)) the failure behaviour will be given by the simple ROM model. Beyond A the initial matrix failure stress will increase substantially with V_f until it coincides with the fibre failure stress (B). Between A and B the composite will continue to sustain load after fibre failure until final fracture at the ROM prediction for fibre failure.

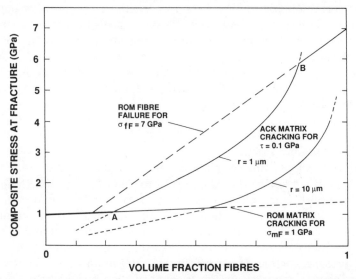

Figure 4.4 Effects of volume fraction and fibre radius on composite strength for simple rule of mixtures and ACK models, with $E_f = 600$ GPa, $E_m = 400$ GPa, $\Gamma = 60$ J/m^2.

The ROM prediction for fibre failure would be correct if all fibres had the same fracture stress. In reality their fracture will be sequential and dictated by their strength distribution. A simple fibre bundle model [7] shows that their failure will yield a non-linear load-extension curve (at constant extension rate the curve can exhibit a maximum). For a Weibull strength distribution the ultimate strength of the bundle is given by

$$\sigma_{bF} = \sigma_0(mL)^{-1/m}e^{-1/m} \qquad (4.11)$$

where L here represents the bundle length which can be approximated to the relaxation length in a composite. σ_{bF} is always lower than the mean fibre strength and so the composite failure is predicted to be somewhat below the ROM line in Figure 4.4.

The above model (i.e. the ACK model combined with a bundle strength model) appears to apply to a good approximation to composites with strong fibres and weak interfaces. In general, however, account must be taken of mechanisms of crack growth and the effects of the stress intensification at the crack tip on the fibre failure and debonding as will be shown in later sections.

4.2.3.2 Short, unidirectional fibres. The key feature of short fibres in a loaded composite is that a finite length at each end of the fibre, l_t, is required for load transfer to the fibre by interfacial shear. Some important consequences of this are that (a) mean stresses and strains are higher in the matrix and lower in the fibres than in the continuous fibre case, (b) stress and strain can be locally very high in the matrix close to the ends of fibres, (c) when matrix cracking occurs, fibres having ends at a distance less than l_t from the crack will be pulled out. Fibres with length less than a critical value $l_c = 2l_t$ will always be pulled out. Again referring to a frictional interfacial bond the transfer length is given by

$$l_t = \sigma_f r/2\tau \qquad (4.12)$$

It should be noted that σ_f and therefore l_t increase with the load on the composite. Consequences of these effects are that as the fibre aspect ratio falls with respect to l_t/r the Young's modulus will fall increasingly below that given by equation (4.6) (see below), the matrix failure strain will fall below that given by the ACK model and fracture determined by fibre failure will become increasingly unlikely.

4.2.3.3 Off-axis fibre orientation. With increasing deviation of the fibre axis from the composite loading axis the axial tensile stress in the fibre falls, shear stresses parallel to the fibre and transverse tensile stresses increase. The matrix and interfacial strengths are usually quite low and therefore the modes of failure discussed above involving fibres bridging the crack are superseded by matrix or interface failure modes. As will be seen later, this does not preclude toughening of the matrix by mechanisms such as crack deflection.

Another consequence of off-axis loading is a marked fall in Young's modulus (the elastic properties of off-axis composites and laminates are discussed in detail in chapter 5).

4.2.3.4 Particulate composites.

Reinforcement particles can be likened to very short fibres with lengths well below the transfer length l_t. Again as will be seen this does not preclude such toughening effects as crack deflection, microcracking and phase transformation.

Several models have been proposed for predicting the elastic behaviour of particulate and short random fibre structures (e.g. see [8–10]). A good approximation is the lower bound model corresponding to a transverse laminate structure:

$$1/E_c = V_m/E_m + V_p/E_p \qquad (4.13)$$

where the subscript p represents the particulate.

4.2.3.5 Multidirectional weaves and laminates.

These structures are employed to ameliorate the poor off-axis properties of unidirectional fibre composites. As already stated their stress–strain and fracture behaviour can be understood qualitatively by analogy to the axial behaviour of unidirectional composites. Off-axis plies and fibre tows behave like a weak matrix giving early cracking while axial plies are analogous to the fibres, providing high ultimate failure stress. The interlaminar shear behaviour acts in analogy to the fibre–matrix interface (see also chapters 5 and 8).

4.2.4 Thermal expansion mismatch and internal residual stresses

Reinforcement and matrix in composites generally have widely different coefficients of thermal expansion. Therefore when a composite is cooled after preparation, thermally induced residual internal stresses develop as a result of the strain constraint imposed by the two phases on each other. An expression for the axial tensile stresses generated in a uniaxial long-fibre composite can be readily obtained as

$$S_m = (\alpha_f - \alpha_m)\, E_f V_f E_m \Delta T / E_c \qquad (4.14a)$$

$$S_f = -S_m V_m / V_f = (\alpha_m - \alpha_f) E_m V_m E_f \Delta T / E_c \qquad (4.14b)$$

where S refers to a thermally induced residual internal stress, α_f and α_m are the coefficients of thermal expansion and ΔT is the change in temperature. Here, ΔT is negative for a falling temperature and a negative stress implies compression.

Mismatch stresses calculated with equation (4.14) for some SiC fibre reinforced ceramics are given in Table 4.1. For short fibres and whiskers these can be assumed to apply close to the centre of the whiskers provided $L > l_c = 2l_t$.

With the exception of composites of silicon nitride and certain low-expansion glasses, SiC reinforced ceramics will exhibit axial compressive stresses in the fibre or whisker and tensile stresses in the matrix following cooling. This results in a corresponding apparent strengthening of the fibre and, more importantly, a weakening of the matrix. Local tensile stresses around a fibre can lead to deflection of the matrix crack paths as well as to the formation of microcracks either during cooling itself or upon subsequent loading. As will be discussed in later sections, deflection and load-induced microcracking can contribute to toughening processes.

While equation (4.14) provides an estimate of the relative level of mismatch stresses, an accurate treatment of thermal mismatch effects requires knowledge of the triaxial stress state around fibres and whiskers. This can be conveniently obtained using numerical approaches. Figure 4.5 gives the results of such an analysis for a specific composite system with long, uniaxial fibres. The predictions of equation (4.14) are confirmed but the analysis also reveals the presence of a small but significant radial stress having the opposite sign to the axial stress in the matrix.

Analysis of mismatch stresses for particulate composites gives similar results. For a single spherical particle in an infinite matrix [12] a hydrostatic stress is created in the particle:

$$P = S_r = S_t = -(\alpha_m - \alpha_p) \frac{\Delta T}{[(1 + v_m)/2E_m] + [(1 - 2v_p)/E_p]}$$

(4.15)

where S_r and S_t are the radial and tangential stress components in the particle. The corresponding stresses in the matrix are then given by

Table 4.1 Estimated thermal expansion mismatch stresses for uniaxial SiC fibre reinforced ceramics: $V_f = 0.2$, $\Delta T = -1000\,°C$ (glass $-800\,°C$)

System matrix–fibre	$\alpha_m - \alpha_f$ ($°C^{-1}$) × 10^6	S_m (GPa)	S_f (GPa)
Al$_2$O$_3$–SiC	4.4	0.48	−1.92
Si$_3$N$_4$–SiC	−1.6	−0.178	0.712
Mullite–SiC	0	0	0
Glass–SiC	5.5	0.22	−0.88

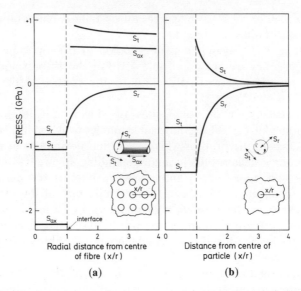

Figure 4.5 Thermal mismatch stresses (a) in a long, parallel fibre composite ($V_f = 0.2$) based on a finite element numerical calculation [11]; (b) for a single spherical particle in an infinite matrix [12]. In both cases the results apply to SiC in an alumina matrix with $\Delta\alpha = 4.4 \times 10^{-6} \text{ K}^{-1}$ and a rigid interface.

$$S_r = -2S_t = P(r/x)^3 \qquad (4.16)$$

where r is the particle radius and x the radial distance from the particle. The results for a SiC particle in an alumina matrix are given in Figure 4.5. In a system with a finite volume fraction of particles the stress inside the particle will be somewhat lower (it can be estimated by replacing the properties of the matrix in equation (4.15) by the mean properties of the composite). However, in the matrix the stress distributions will be a summation of the overlapping stress fields of the particle population. In general, a radial stress of the same sign and magnitude as that in the particle will exist close to the particle but equilibrium dictates that the dominant stress in the matrix must be of opposite sign to that in the particle and at a level proportional to the particle volume fraction (cf. (4.14)).

The use of numerical analysis can be extended to reveal the mismatch stress distribution around the ends of short fibres and whiskers. As expected from analogy with the analysis for an external load discussed in section 4.1.3, a region of interfacial shear stress and falling tensile stress is found at the fibre ends. For example considering the examples in Table 4.1 with rigid interfaces, values of the peak interfacial shear stress around 1000 MPa are predicted. This could well lead to premature debonding on loading if not already upon cooling.

4.2.5 Toughening in long, unidirectional fibre composites

4.2.5.1 Overview of mechanisms. A convenient qualitative description of fracture modes in axially loaded, long-fibre composites has been given by Luh and Evans [13] in terms of a mechanism map based on the important parameters, fibre strength and interfacial shear strength (Figure 4.6). In a region of high fibre strength and low interfacial strength, failure is by complete, multiple cracking in the matrix followed by fibre failure at a subsequently higher composite stress as outlined in section 4.2.3. Recalling Figure 4.4, this corresponds to a region in which the composite fibre-failure stress, $\sigma_{fF}V_f$, is high relative to the matrix failure stress given for example by the ACK equation. Thus the extent of this failure mode region will also depend on volume fraction giving way to fibre initiated failure at both high and low fibre fractions. It has been shown [14, 15] that for large pre-crack or defect sizes in this region the matrix fracture stress is independent of the defect size, that is the material is not sensitive to defects and fibre toughening cannot be described in terms of a fracture toughness parameter such as K_{Ic}.

At higher interfacial strengths and/or lower fibre strengths the fibres will either fail simultaneously with the matrix or at some distance behind the advancing crack tip thus forming a so-called bridging zone (Figure 4.6). An approximate criterion for the transition between multiple matrix fracture and bridging zone formation is $\tau \approx \sigma_{fF}^3 V_f r / E_c \Gamma_m$ [13]. For the fibres to survive the passage of a crack in the matrix and so form a bridging zone some interfacial debonding will normally be necessary. Thus in this expression τ refers to a friction stress. Evans *et al.* [16] have examined the conditions necessary for such debonding and

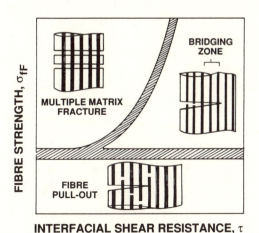

Figure 4.6 Possible fracture mechanisms in uniaxial, long-fibre composites as affected by fibre strength and interfacial bonding (based on [13]).

show that in systems in which fibre and matrix have similar elastic moduli, debonding will occur provided the interface fracture energy is less than a quarter of the fibre fracture energy. In composites with a stronger interfacial bond in which no debonding occurs, the effective fibre strength will usually be severely reduced since it will be determined by the fibre toughness in combination with the matrix crack size. The bridging zone will become insignificant and the composite toughness will be given approximately by a linear mixture of the fibre and matrix toughness.

A third mode of failure, namely fibre pull-out (Figure 4.6) can be expected if the fibres have very low strength and debond readily. This is because they will fracture at some distance from the crack plane in a manner determined by the statistical distribution of defects in them. Bridging fibres that fracture within a distance from the crack plane of less than the shear-lag transfer length (see section 4.2.3) will pull out from the matrix rather than fracture again. The contribution of this pull-out mechanism to toughening has been estimated by Thouless and Evans [17] in terms of its contribution to fracture energy, Γ. In general, toughening is favoured by a low value of τ and a large scatter in fibre strength (large m) since this gives a larger average pull-out length. More details will not be given here; a simpler pull-out model for short fibres is given in section 4.2.6.

4.2.5.2 Models for toughening by fibre bridging. Figure 4.7 shows schematically a zone of bridging fibres adjacent to a crack tip. Distance behind the crack tip is given by x. The zone length is thus x_0 the distance of the last bridging fibre. The crack opening displacement is given by u, u_0 being the opening at the end of the zone. Each fibre is characterised by a debond length. The stress on the fibre, σ_f, is maximum in the crack plane ($\sigma_f = \sigma'_f$) and decreases linearly over the debond length to a value corresponding to the general fibre stress in the undamaged composite.

The toughening produced by such a bridging zone can be estimated in two ways:

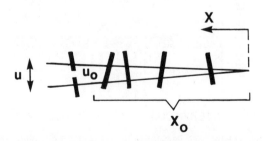

Figure 4.7 Zone of bridging fibres at crack tip.

(i) by evaluating K_I at the crack tip by an integration of the net tractions along the fracture surface [18]:

$$K_I = 2(c/\pi)^{1/2} \int_0^1 \frac{[\sigma_c - \sigma_f(x)]}{\sqrt{1 - x^2}} x \, dx \qquad (4.17)$$

where σ_c is the external stress on the composite and $\sigma_f(x)$ is the stress tending to close the crack exerted by a bridging fibre at a distance x from the crack tip. Equation (4.17) yields the true K_I value, that is a value lower than the apparent value that would be obtained by measurement of σ_c and c. The toughening, ΔK_{Ic}, is the difference between the true and apparent K_I at fracture.

(ii) by estimating the necessary increase in strain energy release rate caused by the fibres [19] through the equation

$$\Delta G = V_f \int_0^{u_0} \sigma_f(u) \, du \qquad (4.18)$$

Note that in discussions of toughening mechanisms both the stress intensity and energy approaches are used. The results can readily be compared through the well-known fracture mechanics relationship for plane strain, mode I fracture:

$$G_{Ic} = K_{Ic}^2 (1 - v^2)/E = \Gamma_I \qquad (4.19)$$

where G_{Ic} is the critical strain energy release rate required to overcome the fracture surface energy Γ. E in the present context refers to E_c.

Both methods should lead to equivalent results; both require knowledge of the fibre stress as a function of distance behind the crack front or of crack opening which in turn depends on the interfacial bonding. The nature of the relationship between stress and distance has a significant effect on the toughening but is rarely easy to derive. Here two examples will be given

Bridging model 1: For debonded fibres in which debond length increases linearly with distance behind the crack tip, Evans and McMeeking [19] give

$$\sigma_f' = (4\tau u E_f/r)^{1/2} \qquad (4.20)$$

Substituting this into equation (4.16) gives

$$\Delta G_c = V_f \, \sigma_{fF}^3 \, r/6E_f\tau \qquad (4.21)$$

Bridging model 2: Becher *et al*. [20] have given an analysis based on (4.17) with the assumption that for a given composite the debond length is constant and given by

$$l_d \approx r\Gamma_m/6\Gamma_i \qquad (4.22)$$

where Γ_i is the fracture surface energy of the interface. This leads to a toughening expression:

$$\Delta K_c = \sigma_{fF} \left[V_f r E_c \Gamma_m / 6 (1 - v^2) E_f \Gamma_i \right]^{1/2} \tag{4.23}$$

For large toughening increments equations (4.21) and (4.23) have the same functional form; thus through equation (4.19), (4.21) becomes

$$\Delta K_c = \sigma_{fF} \left[V_f r E_c \sigma_{fF} / 6 (1 - v^2) E_f \tau \right]^{1/2} \tag{4.24}$$

Both models reveal how toughening is increased by increasing fibre content, strength and radius and by decreasing interfacial strength and fibre modulus. Inserting appropriate values into these equations yields values of toughening consistent with experimental observations. For example 20 vol% of a 10 μm diameter fibre having $E_f = 300$ GPa, $\sigma_{fF} = 2000$ MPa into a matrix of similar stiffness and an interfacial friction stress of 10 MPa would according to equation (4.24) give a toughening increment of about 10 MN m$^{-3/2}$ implying a debond length of about 150 μm in equation (4.23).

Both models imply a steady state condition, that is as the crack extends through the material the bridging zone remains constant. An initiating flaw however will not normally have a developed zone. Thus from the start of fracture until the steady state is reached a rising toughness, that is R-curve behaviour will be experienced.

4.2.6   Toughening in short-fibre and whisker reinforced composites

Four important mechanisms of whisker toughening that have been proposed and observed experimentally are illustrated schematically in Figure 4.8, namely bridging, pull-out, deflection and microcracking. These will be discussed in turn below. A further mechanism, crack branching is not treated explicitly but is considered in many respects to be similar to deflection.

4.2.6.1   Bridging.   The principles of bridging discussed in the previous section apply to short fibres and whiskers. A rigorous model should however where necessary take account of fibre orientation and allow for a contribution of fibre pull-out for a proportion of fibres having ends close to the fracture plane.

Insertion of data appropriate to whisker reinforced ceramics into equation (4.21) or (4.24) predicts toughening increments of up to around 100 MN m$^{-3/2}$. Such toughening is not observed in practice no doubt because many of the whiskers are too short or misoriented. However, such an estimate indicates the potential of this mechanism for the very high values of σ_{fF} achieved in whiskers.

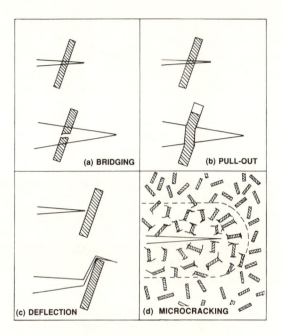

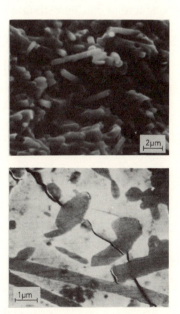

Figure 4.8 Some proposed toughening mechanisms in short-fibre reinforced ceramics (note that pull-out of a fibre inclined to the crack plane must involve fibre bending [6]). The scanning electron micrographs show a fracture surface and stable crack profile in SiC-reinforced alumina illustrating bridging, pull-out and deflection.

Inserting more realistic values of debond length into equation (4.23) yields toughening values in excellent agreement with experiment [20]. This to some extent may be fortuitous since the model takes no account of pull-out and deflection. However, these mechanisms are in many respects similar in nature and could to some extent be interchangeable.

A further implication of equation (4.23) is that the bridging zone is around 10 μm in length which is in accord with experimental observations. A result of this is that R-curve behaviour is expected to be very limited.

4.2.6.2 Pull-out.

As already indicated a condition for pull-out is that fibre ends come within the stress-transfer length before the fibre stress reaches σ_{fF}. The effect of fibre orientation is not obvious; misorientation will reduce the axial stress in the fibre but also induces bending thus reducing the effective strength. The model considered here [8] is simplified by considering a system of fully debonded fibres, aligned uniaxially and normal to the crack plane (Figure 4.9). They are assumed to have a uniform length, l_f, that is less than l_c ($= 2l_t$), the critical length for fracture; that is they will all be pulled out. At an intermediate stage of pull-out the fibre will experience a maximum stress, σ_f', between the crack faces. If σ_f' can be derived in terms of the crack opening, u, then the toughening can be estimated with equation (4.18). σ_f' is determined by the shear transfer on the length of fibre still not pulled out, that is it lies between 0 for complete pull-out and

$$\sigma_f'(\max) = (2\tau/r)[(l_f/2) - u] \tag{4.25}$$

for fibres that are divided equally by the fracture plane. Since σ_f' is directly proportional to the embedded length (τ being constant) and the lengths are uniformly distributed between 0 and ($l_f/2 - u$), then for a given distance behind the crack tip (i.e. a given value of u) the average value of σ_f' is

$$\sigma_f'(\mathrm{av}) = \sigma_f'(\max)/2 = (\tau/r)[l_f/2) - u] \tag{4.26}$$

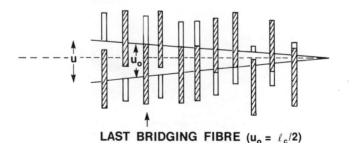

LAST BRIDGING FIBRE ($u_o = \ell_c/2$)

Figure 4.9 Fibre pull-out across crack-tip region.

The effective volume fraction of fibres involved decreases in proportion to u since increasing numbers of fibres become completely pulled out. Thus

$$V_f(u)_{\text{eff}} = [(l_f/2) - u]/(l_f/2) \tag{4.27}$$

and (4.18) becomes

$$\Delta G_c = (\tau V_f)/(2rl_f) \int_0^{u_0} (l_f - 2u)^2 \, du \tag{4.28}$$

Integrating and setting $u_0 = l_f/2$ gives

$$\Delta G_c = \tau V_f \, l_f^2/6r \quad (l_f < l_c) \tag{4.29}$$

In the special case of fibres having the critical length, remembering that $l_c = \sigma_{fF} r/\tau$,

$$\Delta G_c = V_f \sigma_{fF}^2 r/6\tau \quad (l_f = l_c) \tag{4.30}$$

Since l_c represents the longest fibre that can be pulled out, equation (4.28) represents the maximum possible toughening by pull-out. With increasing length above l_c, whiskers will be increasingly less likely to pull out. This equation indicates the apparent benefits of high fibre strength and radius as well as low interfacial friction. However, it must be borne in mind that both (4.29) and (4.30) apply to fully developed bridging zones of lengths corresponding to a crack-opening of $u_0 = l_f/2$. High values of ΔG imply high transfer lengths and consequently unrealistically high values of crack opening. In practice, R-curve behaviour can be expected in which the ultimate toughening is achieved only after some crack growth. For long fibres this will only be attained for very long cracks.

As an example, parameters for the system considered above for whisker bridging ($\tau = 10$ MPa, $V_f = 0.2$, $r = 0.5\,\mu$m) can be inserted with a value of $l_f = 10\,\mu$m into (4.29). This gives a value of $\Delta G_c = 67$ J/m^2 which corresponds to an increase in K_{Ic} from 4 to 7 MN m$^{-3/2}$ in this material; that is a similar magnitude to bridging toughening. To attain this ultimate value of toughening would require a crack opening of $u_0 = 5\,\mu$m.

For whiskers of strength, $\sigma_{fF} = 10$ GPa the critical length would be $500\,\mu$m. A composite with aligned whiskers of this length would according to equation (4.28) achieve a toughening of $\Delta G_c = 166$ kJ/m^2 corresponding to a K_{Ic} of around 280 MN m$^{-3/2}$ but requiring a crack opening of $250\,\mu$m. Equation (4.29) is evidently more realistic for currently available whisker reinforced ceramics. The toughening contribution is of a similar magnitude to that of bridging and exhibits a similar dependence on whisker fraction. Thus to some extent these two processes can be considered to complement each other when estimating the effects of whisker content on the level of toughening.

4.2.6.3 Crack deflection. Crack deflection has been analysed in detail by Faber and Evans [21]. A crack can be expected to deflect around a whisker if the whisker exhibits an appropriate combination of inclination to the crack plane and relative interfacial strength such as to prevent fibre fracture or fibre pull-out. Deflection will also be encouraged by residual mismatch stresses around the whisker (section 4.2.4). The local crack deflection will vary between tilt and twist depending on the orientation (Figure 4.10). Each local segment can then be considered to contribute strain energy release rates appropriate to its relative components of local mode I, II and III fracture. The Faber and Evans analysis outlined briefly below (and hereafter abbreviated FE) shows that a summation of these local release rates leads to a reduced driving force relative to an undeflected crack front having the same forward growth.

$$K_I(\text{tilt}) = K_{11} K_I \qquad (4.31a)$$

$$K_{II}(\text{tilt}) = K_{21} K_I \qquad (4.31b)$$

where K_{11} and K_{21} are functions of the angle of tilt. Similarly a length of twisted crack at the tip experiences components of mode I and mode III stress intensity expressed by similar equations as a function of twist angle [21]. To assess the effects of these local deviations on toughening, FE propose a summation of the local contributions of the three modes to give an effective virtual strain energy release rate, G_T, using the general form of equation (4.19). Then by simple proportionality the critical strain energy release rate (i.e. toughness) of the composite is given by

$$G_c = (G_u/G_T) G_c' \qquad (4.32)$$

where G_u is the virtual strain energy release rate for the undeflected crack and G_c' is the toughness of the non-deflecting material.

From Figure 4.10 it should be apparent that the degree of combined tilt and twist between two neighbouring whiskers is dependent on their angular orientation (defined by two angles, each), their distance above or below the undeflected crack plane and their distance apart as well as

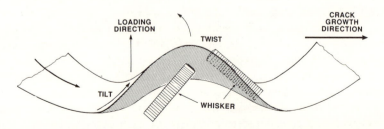

Figure 4.10 Illustrating schematically tilt and twist in crack deflection.

on their aspect ratio. FE derive an expression for G_T for given volume fraction and aspect ratio of randomly distributed rods involving integration over the four angles of orientation and the two relative distances from the undeflected plane. This is extremely involved, requiring numerical integration, and is not reproduced here. The results of the integration for various aspect ratios and volume fractions are given by FE. These indicate that the toughening effect tends towards saturation at V_f around 0.2 and aspect ratio around 10, giving an upper limit for G_c/G_c' of around 4. These results of FE can be converted back to be expressed in terms of effective K_{Ic} via (4.19):

$$K_{Ic}/K_{Ic}' = [(G_c/G_c')(E(1 - v'^2)/E'(1 - v^2)]^{1/2} \qquad (4.33)$$

This has been done in Figure 4.11 for a composite with $E' = E_m = 400$, $E_f = 600$ and $v' = v = 0.3$. For most whisker reinforced ceramics $E \approx E'$ implying a toughening factor limit of about $2 \times$.

FE also demonstrate that the degree of toughening to be expected merely from the increase in crack surface area associated with deflection is significantly lower than that due to the entire reduction of strain energy release.

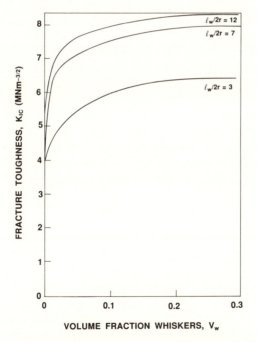

Figure 4.11 Crack deflection toughening in a whisker reinforced ceramic with $E_m = 400$ GPa, $E_f = 600$ GPa and a matrix toughness of 4 MN m$^{-3/2}$ (based on results of [21]).

4.2.6.4 Microcracking. Microcracking as a means of toughening in ceramics has been analysed in detail by Hutchinson [22]. To be effective the microcracking must occur only in response to the stress field around the crack tip in conjunction with residual stresses and be restricted to small well-dispersed sites (to avoid microcrack linkage). Thus microcracks formed throughout the microstructure, for example during processing, are of less value. The formation of a zone of microcracks ahead of the crack makes two contributions to toughening; firstly, through the formation of a zone of lower elastic modulus and secondly, through the absorption of strain release energy (this can also be thought of as a crack closing effect resulting from the dilation caused by the microcracking in conjunction with residual stresses). Short fibres can provide a suitable source of microcrack sites, for example through tensile failure of the interface at the end faces, through shear failure in the transfer length, or by cracking of the matrix encouraged by thermal mismatch.

The microcracking considered here is that which occurs only when a certain critical stress σ_m is attained in the microstructure. Thus at the tip of a loaded crack a microcracked zone will develop bounded by this stress level (Figure 4.12). The size of the zone will depend on the stress intensity factor, K, while its shape will depend on the form of the stress distribution around the tip and the details of the microcracking mechanism. Within the zone, the microcracking will relieve stress and the density of microcracks will therefore tend towards a saturated value. The reduction of elastic moduli and a permanent expansion of the material in the zone both contribute to a reduction in the true stress intensity, that is to a toughening which will normally outweigh any loss in intrinsic toughness of the matrix that might be caused by the cracking. If the crack grows, the zone will remain as a microcracked zone of constant thickness (h) in the wake of the crack tip (Figure 4.12) where it will continue to contribute to toughening and therefore R-curve behaviour is expected to occur from crack growth initiation up to a steady-state growth state for a large wake length.

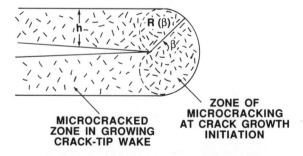

Figure 4.12 Microcrack zone around a crack tip.

Hutchinson [22] analysed this type of microcrack toughening for various modes of microcracking both for the crack initiation and steady-state wake situation. In general, the reduced stress intensity at the crack tip caused by microcracking can be expressed as

$$K_m = K' - A v K' - BE\delta\sqrt{h} \qquad (4.34)$$

where K' is the stress intensity in the absence of microcracking. The term Av represents the toughening effect of elastic moduli reduction, v being a measure of the microcrack density (for penny-shaped cracks, $v = Nb^3$ where N is the number of cracks per unit volume and b is the crack radius) and A is a coefficient depending on the shape of the zone (i.e. $R(\beta)$) and the relative reduction of moduli (which itself depends on v). The last term in equation (4.34) represents the contribution of the zone dilation to toughening; B is a function of the zone shape and the poisson ratio, v; h is defined in Figure 4.12 and δ is the fractional dilation which for penny-shaped microcracks is given by

$$\delta = 16(1 - v^2)vS_r/3E \qquad (4.35)$$

S_r being the residual stress normal to the microcrack plane.

To apply Hutchinson's model to fibre- and whisker-containing microstructures requires knowledge of the details of microcracking for each particular case. For purposes of illustration here, it will be assumed that each fibre forms two penny-shaped microcracks but only when the maximum principal stress at the microcrack site reaches a critical value, σ_{cr}, a criterion considered by Hutchinson. For this critical stress criterion with isotropic microcrack orientation, Hutchinson gives the value of A and B as 1.15 and 0.16 for the case of crack growth initiation (i.e. no wake) and 1.42 and 0.39 for a steady-state crack wake. The zone height h is given by $h \approx 0.25 (K'/\sigma_{cr})^2$.

Assuming the microcrack radius, b, to be approximately $2r$ (half the fibre spacing) then the value of v is given by

$$v = 8nV_f r/\pi l_f \qquad (4.36)$$

where n is the number of microcracks forming per fibre. This indicates a benefit of short fibres. With the above conditions putting $n = 2$ and as in previous examples putting $r = 0.5 \ \mu m$, $l_f = 10 \ \mu m$, $V_f = 0.2$, $K' = K_{Ic} = 4 \ MN \ m^{-3/2}$, $v = 0.3$ and for convenience letting $\sigma_m = S_r = \sigma_{cr}$ it is found that $K_m = 3.69$ for crack growth initiation and 3.52 at steady state, implying toughening of 8 and 13.5%, respectively. This rough estimate indicates that significant toughening might feasibly be achieved through microcracking. Greater toughening increases would for example be obtained by increasing the number of microcracks forming at each whisker or by forming larger cracks.

4.2.6.5 Conclusions. The above treatments show that the proposed mechanisms of whisker toughening can under certain circumstances be expected to give significant toughening of ceramic matrices. Of particular benefit is the fact that the different mechanisms are additive and can to some extent be expected to complement each other. For example, a whisker unfavourably oriented for pull-out is likely to be favourably oriented for deflection.

4.2.7 *Toughening in particulate composites*

4.2.7.1 Overview. Four toughening mechanisms identified in particulate ceramic composites are deflection, microcracking, transformation toughening and matrix compression. Mechanisms of deflection toughening are analogous to those discussed for short-fibre composites and will not be considered specifically here. For particles having an aspect ratio close to 1 the toughening is predicted to be somewhat less than for fibres (see Figure 4.11). Microcracking can be induced in particulate composites as a result of thermal mismatch stresses exactly as discussed above for whiskers and again the analysis will not be repeated. A more effective source of microcracking, however, can be the internal stresses resulting from a change in particle volume when it undergoes a phase transformation. Since this effect is closely associated with transformation toughening, the two mechanisms will be treated following a general discussion of the transformation phenomenon below.

The residual matrix compression present in composites in which $\alpha_p > \alpha_m$ has been proposed as a possible source of toughening [23]. Provided the particles are sufficiently strong, the fracture path will pass through the matrix which will have an increased effective toughness due to the compression state.

Bridging and pull-out are not expected to be significant mechanisms in particulate composites. A possible exception is toughening caused by ductile metallic inclusions since these can extend plastically across a crack and contribute significant closure tractions and energy dissipation [24]. Such composites are not within the scope of this book.

4.2.7.2 Transformation toughening. Toughening can be produced in particulate composites if an added particulate constituent undergoes a phase transformation involving a volume increase. A number of ceramic compounds exhibit phase changes and are potentially exploitable for transformation toughening [25]. By far the most common example is zirconia toughened alumina (ZTA) and this will be used to exemplify the present treatment.

The phases present in zirconia are illustated in Figure 4.13 in terms of

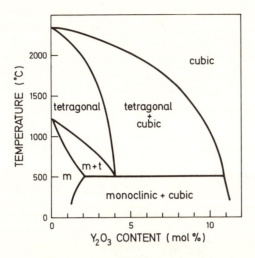

Figure 4.13 Pseudo-binary phase diagram between zirconia and yttria.

a binary phase diagram between zirconia and yttria. Here the effect of yttria is representative of a number of oxides which act to stabilise the higher temperature form of zirconia. The highest temperature form is cubic. In pure zirconia this transforms to the tetragonal form below about 1050 °C. The effect of a stabilising oxide is to reduce the transformation temperatures and to create two-phase intervals (i.e. $c + t$ and $t + m$). Of special significance is the fact that the t m transformation is martensitic, that is it occurs by a diffusionless process (twinning). Thus the transformation is not thermally activated but occurs when the temperature is low enough to provide an excess free energy of transformation sufficient to overcome possible nucleation barriers and to compensate for twin boundary and accommodation strain energy. The basis of transformation toughening with zirconia is that the transformation involves a volume increase ($\approx 4\%$) and is stress-assisted, for example by the stresses generated by growing defects and cracks. The transformation leads to two distinct mechanisms of toughening: (i) the transformation itself having occurred at the tip of a growing crack exerts closing forces on the crack; (ii) the transformation occurs prior to crack growth but the transformation strain can be the source of microcracking. If the microcracking is activated by a growing crack then effective toughening is achieved as discussed in section 4.2.6.

The transformation contribution to toughening has been treated analytically by a number of authors, the main developments having been recently reviewed by Evans [26]. The main features of transformation toughening models for ZTA are illustrated in Figure 4.14. It is assumed that a steady state transformation zone of half-width h develops around

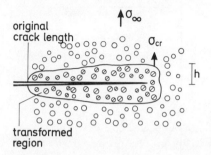

Figure 4.14 Zone of transformed particles in a ceramic matrix, surrounding a growing crack indicating factors involved in transformation toughening.

a growing crack. In the simplest case it is assumed that the fraction of transformed particles, V_{pT} is constant over the whole zone and that the boundary of the zone is set by the attainment of a critical transformation stress, σ_{cT}. The increase in effective toughness (equivalent to a reduction in stress intensity at the crack tip due to closure) is given by [26]

$$\Delta K_c = A E V_{pT} \varepsilon_T h^{1/2}/(1 - v) \qquad (4.37)$$

where E is the elastic modulus of the composite and ε_T the transformation strain. A is a numerical constant dependent on the details of the transformation mechanism. If the transformation is activated by the hydrostatic stress around the crack tip then $A = 0.22$. If it is activated by shear then values of A up to 0.38 are derived [26, 27]. The value of h is clearly an important variable; it is determined largely by the relative values of the local stress level and the transformation stress:

$$h = \sqrt{3}\, (1 + v)^2 (K_{cm}/\sigma_{cT})^2/12\pi \qquad (4.38)$$

where K_{cm} is the toughness of the matrix. From the start of crack growth until the steady state is reached, R-curve behaviour is expected. Under the conditions of the above model the steady state is reached when the crack growth $\Delta a \approx 5\,h$.

A practical implication of the above model is that σ_{cT} should be as low as possible (but not so low as to risk transformation prior to crack growth). To achieve this the martensite transformation temperature should be adjusted to be a little below the expected service temperature. The main factors that affect the transformation of a zirconia particle are: (a) the constraint of the surrounding material—high modulus and high thermal expansion coefficient suppress transformation; (b) particle size and shape—the transformation is suppressed by decreasing particle size since this hinders nucleation consistent with classical nucleation theory

[27]; angular particle shapes, as are found in particles formed at matrix grain-boundary intersections, encourage transformation since higher stresses are developed locally in non-spherical particles; (c) stabilisation by dissolved oxides (see above).

In general, the degree of toughening is expected to pass through a maximum with changing particle size or with solute content as the transformation temperature changes. The fact that the composite is optimised for a given temperature implies that the degree of toughening will fall with increasing temperature. Evans and Cannon [27] have presented a detailed review of nucleation and transformation mechanisms; these are complex and difficult to quantify with precision. In practice, the optimum particle size lies around 1 μm.

The model presented above predicts that the degree of toughening should increase continuously with the volume fraction. In practice there is an upper limit to this effect and in ZTA composites the toughness is often found to pass through a maximum somewhere between 10 and 30 vol% (Figure 4.15). Becher has quantified this effect in terms of a reduction in the constraint of the material surrounding the particle leading to partial transformation of particles during processing [28]. Other factors may include the development of particle contact and an increase in microcracking. Another effect not accounted for by the above model is the influence of the transformed zone on the continued transformation in the material ahead of the tip. The zone itself will give rise to tensile stress field in the surrounding region thus encouraging the development of an even increasing transformation zone width.

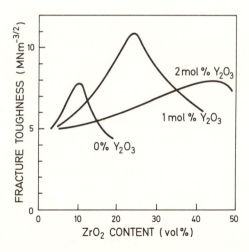

Figure 4.15 Effects of volume fraction zirconia particles and level of stabilisation on toughening in ZTA (after Becher [28]).

4.2.8   Combined toughening mechanisms

4.2.8.1   Multiple toughening and synergism.   Needless to say, two or more of the above toughening mechanisms can operate simultaneously [29, 30]. As an example, Kageyama and Chou [29] have considered a combination of crack deflection and pull-out in whisker reinforced composites, the degree of deflection and proportion of pull-out being determined on a statistical basis. Even if the contributions of different mechanisms can be treated simply as additive, a successful general description of the toughening must be capable of predicting the relative contribution of each. This subject is treated later in this section.

Of at least equal interest is the combination and interaction of short fibre and whisker toughening with other toughening mechanisms. Such multiple toughening is not necessarily simply additive since depending on circumstances the presence of whiskers might either reduce or enhance another toughening process. An important example is the combination of whisker toughening with zirconia-based transformation toughening. Of significance in the present context is that the degree of toughening in a ceramic material containing zirconia particles is related to the size of the transformed zone which increases with the stress level at the crack tip, that is with the K_{Ic} value of the matrix. Thus, an independent toughening of the matrix with whiskers would be expected to enhance the transformation toughening and thereby produce an increase in toughness greater than that given by a simple summation of the two [31]. Analogous reasoning applies in cases where toughening is produced by a zone of stress-induced microcracks.

4.2.8.2   Predicting combined whisker toughening mechanisms.   As was shown in sections 4.2.5–4.2.7, quantitative predictive models exist for most of the proposed toughening mechanisms and these can be combined arbitrarily as indicated in the previous section. However, criteria for predicting under which conditions a given mechanism will operate are less well developed. Thus, although progress has been made recently in predicting mechanism transitions [16, 26, 32], a generalised model yielding an unconditional prediction of toughening as a function of microstructure is still unavailable. The nature of the problem can be appreciated by listing the choices that face a growing crack as it meets or passes a second phase particle or fibre and the parameters that influence the choice:

(1) The crack meets the fibre and this fractures without significant debonding. This response is favoured by strong fibre/matrix bonding, low fibre fracture strength, σ_{fF}, and high values of θ, the angle of the whisker axis with respect to the crack plane.

(2) Either the fibre is already debonded or debonds as the crack front approaches; the crack continues past the fibre to leave a bridging fibre; debonding may continue and eventually the fibre fractures in the crack wake. This mode is favoured by moderate bonding, moderate σ_{fF}, long fibres (embedded length $> l_c/2$), high θ.

(3) Similar to (2) but the fibre pulls out on one side of the crack instead of failing; favoured by weak interface (much debonding), high σ_{fF} and short fibres (embedded length $< l_c/2$), high θ (see Figure 4.8b).

(4) The approaching crack front deflects around a fibre or is deflected towards or away from a fibre close to the main crack plane; favoured by weak interface, high thermal expansion mismatch, low θ. For higher θ values a high σ_{fF} is necessary to avoid the first mechanism, that is direct fibre fracture. However, in combination with low θ fibre fracture could itself lead to a form of deflection.

(5) The passing crack causes isolated microcracks to form at or in surrounding fibres. Microcracks forming ahead of the crack tip close to the crack plane may link up with each other and the main crack and thereby be involved in the crack growth process, for example they might encourage deflection; favoured by low interface strength, high thermal expansion mismatch.

Any general model will have to consider all fracture modes together taking into account all the above parameters. The first step, that is the development of an adequate description of the influence of a crack tip on both interfacial debonding and fibre fracture has been taken [16, 32].

4.2.9 Strength variablity in composites

As has been shown the incorporation of second phases in a ceramic can be expected to increase the effective toughness, K_c (or G_c). Assuming the microstructure and in particular the defect distribution remains otherwise unchanged, this would be expected to raise the level of the strength distribution, σ_0, but not the value of the Weibull modulus m (see section 4.2.2). However, closer examination shows that if the incorporation of a second phase causes or intensifies already existing R-curve behaviour then a significant reduction in strength scatter, that is increases in m, can be expected. This effect has been analysed in detail by Cooke and Clarke [33] who give a general expression for the dependence of m not only on the shape of the R-curve but also on possible residual stresses around the defect.

Qualitatively, the effect can be understood on the basis of the fact that an R-curve toughening implies that defects will generally experience some stable crack growth prior to final catastrophic failure. The criterion for failure is that instability occurs when the rate of increase of stress intensity factor with the crack growth, $dK_i/d\Delta c$ (which increases with increasing applied stress) attains the value of the slope of the rising R-curve, $dK_c/d\Delta c$ [1, 34]. The term $dK_i/d\Delta c$ depends not only on the applied stress but also on the original defect or pre-crack size c_0 such that there occurs an increase of apparent fracture toughness with c_0. Consequently, when R-curve behaviour is introduced or intensified in a ceramic, the larger defects will experience a greater increase in failure stress than smaller defects thus causing a narrowing of the strength distribution. In favourable circumstances, order of magnitude increases in m are predicted [33]. It is particularly beneficial if the R-curve is steep in the range of Δc applicable to the critical defect size range in the material.

A second effect of second phase features might be the introduction of a population of larger but more uniform defects which would have the effect of cutting back the upper tail of the strength distribution and thereby also reducing m.

4.3   Time-dependent failure processes

4.3.1   *Sub-critical crack growth*

It has been shown that, in certain circumstances, brittle ceramics can exhibit stable crack growth even under conditions of continuously increasing load or strain, namely when the material exhibits R-curve behaviour or if the component geometry or applied stress distribution are such as to cause a decreasing stress intensity factor with crack growth. Another form of stable crack growth is sub-critical crack growth which occurs at stress intensities below the critical value and involves time-dependent processes occurring at the crack tip. The most important examples are growth under cyclic loading (fatigue), growth involving chemical interactions with the environment [18] and high-temperature fracture involving creep processes at the crack tip. Although these phenomena have been treated extensively, both experimentally and theoretically, for ceramics generally, no detailed analyses have yet been made for ceramic composites. For this reason the present treatment will be restricted to brief qualitative presentations of high-temperature failure and fatigue crack growth with indications of the possible effects of second phase particles and fibres.

4.3.2 *High-temperature failure and creep*

In general, ceramics remain brittle up to relatively high temperatures. However, most polycrystalline ceramic materials exhibit a relatively distinct temperature range in which the defect-dominated brittle behaviour gives way to failure processes determined by time-dependent deformation behaviour. Up to the transition, the fracture toughness and consequently the fracture strength does not normally change significantly (exceptions are among those ceramics that exploit transformation toughening). At the onset of the transition a limited process zone may form at the crack tip giving an intermediate increase in fracture toughness but above the transition the deformation processes which lead to failure usually initiate at significantly reduced stress levels. Thus the strength generally falls as well as becoming strongly strain-rate dependent. These effects are discussed further below.

In pure ceramics the softening processes referred to above are in most cases diffusion-related although in certain oxide ceramics dislocation processes have also been demonstrated [35] while in glasses viscous flow occurs. In fine-grained materials, grain-boundary sliding may be a dominant process. The softening transition is particularly marked in ceramics containing softer phases at grain boundaries such as sintering additives. This has been clearly demonstrated in a study of the tensile behaviour of a hot pressed silicon nitride containing yttria and alumina as additives [36]. In such materials fracture initiates and develops by the formation of voids at the grain boundaries.

Considering firstly conditions of continuously increasing load or strain (i.e. a fracture strength or fracture toughness test), the fracture behaviour depends very much on the strain rate and how far the temperature lies above the transition. At fast testing rates a limited process zone may form at the pre-crack or dominant defect thereby raising the fracture toughness. At somewhat slower rates/higher temperatures, controlled sub-critical crack growth will initiate from the controlling defect/pre-crack, the criterion being that a process such as grain-boundary void formation and linkage is a sufficiently fast process to maintain the necessary crack opening. As the crack grows there may occur a transition to unstable fracture since the effective stress intensity increases with crack length. With decreasing strain rates/higher temperatures this becomes increasingly less likely since there is a simultaneous increase in crack blunting. Interpretation of the fracture behaviour using limited yield fracture mechanics is no longer feasible. At even higher temperature and slower strain rates blunting processes and general deformation become so extreme that growth of a single crack ceases. The material becomes defect insensitive and failure is by general rupture.

Failure at very low strain rates or under constant load conditions, that is creep, follows a similar pattern to the above. It is found that for a given temperature there is a transition stress level above which creep failure can occur by sub-critical growth of a single crack and below which the material is defect insensitive, failure occurring by general creep and void generation throughout the material [37, 38]. In the latter regime the creep mechanism in ceramics are generally diffusion-based with, in fine-grained ceramics, associated grain-boundary sliding. Here the creep rate can often be described conveniently by the equation

$$\dot{\varepsilon} = \text{const } \sigma^n \exp\left(- Q/RT\right) \tag{4.39}$$

where n is a constant usually lying between 1 and 2 and Q is the activation energy of the controlling process.

Although no detailed analyses have been made of the effect of second phase on high-temperature failure certain qualitative predictions can be proposed:

- *Load transfer:* it was concluded in section 4.2.3 that load transfer is expected to contribute little to the strengthening of brittle ceramics. However, beyond the softening transition some effect in inhibiting deformation processes may be expected in those cases where the deformation stress levels are low [39].
- *Adapted toughening mechanisms:* phenomena such as whisker bridging and deflection can be expected to inhibit the initiation and growth of the sub-critical crack.
- *Barrier effect:* it can be supposed that appropriately oriented whiskers would provide a barrier to certain shear deformation processes such as grain-boundary sliding [40].

That whiskers do in fact improve creep resistance has been confirmed for SiC reinforced alumina [40–42].

4.3.3   *Fatigue*

Cyclic-loading fatigue failure occurs when a material is subjected to a varying range of stress below the fracture stress. This can occur in ceramics that exhibit non-linear elastic fracture behaviour with some form of non-reversible change occurring in the microstructure around defects or the growing crack tip. For example, Lewis [43] has suggested that in fibre reinforced ceramics a crack-bridging fibre partially pulled out at the peak stress could be crushed during the subsequent drop in stress. The load-bearing capacity of the sample would then be lower in the next cycle.

Suresh and co-workers [44–46] have shown that ceramics (including whisker reinforced ceramics) are frequently susceptible to cyclic fatigue failure in compression/compression loading even at very low stresses relative to the compressive strength. In general an axial compressive cyclic load leads to the growth of a plane, mode I crack normal to the loading axis. This is clearly a serious weakness in materials intended for engineering applications. It is found experimentally [46] that a frequent cause of this process is the formation of microcracks in a zone around the crack tip during the loading branch of the cycle. This zone then creates a residual tensile stress upon unloading. Above a threshold compressive stress level, this residual stress is sufficient to cause crack extension. Suresh and Brockenbrough [45] have developed constitutive predictive models for this type of fatigue failure for the case of microcracking and of ceramics containing particles of tetragonal zirconia that undergo the transformation to the monoclinic form. The model predicts successfully the crack growth rate and direction as well as load dependence. An implication of this analysis is that the presence of whiskers or fibres is unlikely to improve the fatigue resistance.

4.3.4 *Failure by thermal stresses*

A distinction can be drawn between two types of internal stress in composites caused by temperature changes. The first is that caused by thermal expansion mismatch between the constituent phases and already discussed in section 4.2.4. The second is caused by local temperature variations which also cause expansion mismatch, this time in response to different temperature. Both can give rise to sufficiently high stresses to cause internal cracking which degrades the mechanical properties. If the damage involves any irreversible changes then the material may be susceptible to accumulation of damage during successive temperature cycles, that is thermal fatigue. Thermal fatigue in ceramic composites has not yet received much attention but as has already been noted certain composites exhibit irreversible microstructural processes and thus could be susceptible to this form of failure.

The phenomenon of thermal shock which is failure caused by a sudden change in local temperature and which is a more immediate problem has received some attention. The subject has been reviewed in relation to composites by Hasselman [47]. Consider a sample at high temperature quenched to a lower temperature. Then the temperature gradient set up in the material will depend partly on the external details of the experiment such as the temperature change, ΔT, the quenching rate and the sample geometry and partly on the thermal conductivity of the material. For a given gradient the stress developed will increase with expansion coefficient and the elastic modulus. The resistance to damage by this stress will increase with the strength or toughness of the material.

It can be understood that the phenomena is complex and as yet it has not been possible to describe analytically. Instead use is made of 'figures of merit' [47] which are intended to predict the relative resistance of a material to thermal shock. An example based on the above principles and used in the present work is

$$\Phi = (1 - \nu)KK_{Ic}/E\alpha \qquad (4.40)$$

where K is the thermal conductivity. Such parameters are clearly only rough guides since they do not attempt to indicate the relative importance of the various properties. Experimentally, thermal shock resistance is measured by measuring strength degradation as a function of ΔT in a quenching experiment.

Trends in the thermal shock resistance of composites can be predicted through the parameters making up the figure of merit, for example Φ. Thus the resistance can be improved choosing composite constituents that improve toughness and thermal conductivity but reduce the elastic modulus and expansion coefficient. The thermal conductivity and expansion of composites are discussed in section 7.

In addition to such an optimisation of the property profile, Hasselman [47] has shown both experimentally and with a fracture mechanics model that a significant improvement in thermal shock resistance (i.e. the ΔT required for a sharp strength degradation) can be achieved by the introduction of a population of microcracks with controlled size and number. This generally results in a reduction in initial strength but improves shock resistance by reducing α and E and by limiting additional crack growth. The method is particularly effective if the cracks exhibit R-curve behaviour. As seen in earlier sections composite principles can be used to achieve microstructures with appropriate controlled microcrack populations.

4.4   Indentation and compression

4.4.1   *Indentation*

Perhaps the most well known form of indentation is the hardness test which measures the resistance of a material to permanent indentation damage. However, indentation has much wider significance than this, not least for ceramic materials. The permanent damage occurring in a ceramic under a sharp, hard indentor is usually a complex combination of plastic deformation and microcracking. In addition larger cracks are usually created outside of the contact area. Under blunter indentors, for example spheres or flat punches, the deformation may remain elastic until the load is sufficient to cause cracking outside of the contact area. Thus indentation has been used in a variety of configurations not only to measure hardness but also to study the development of contact damage, to measure the resistance to fracture and to create controlled, stable cracks. In composites indentation has also been applied to individual fibres to assess interfacial bonding [48]. Recently an instrument known as the 'nanoindentor' has been developed that indents at very low loads permitting the *in situ* investigation of the individual constituents of for example composites at a very fine level [49]. Such aspects of indentation relevant to composites are discussed in the following.

4.4.1.1 Hardness. The hardness of a ceramic material is convenient to measure and is one of a number of parameters of relevance in wear and erosion processes (see section 4.5). The most common test applied is the Vickers diamond test which yields the Vickers hardness number (for a review of the hardness of ceramics see e.g. [50]). The complexity of the impression deformation means that the hardness number cannot be directly related to a single, well-defined mechanical process. However it has been shown experimentally that the hardness of composites can be described by simple rules of mixtures in a similar way to the elastic modulus. For example, the hardness of particulate composites lies somewhere between the bounds of the parallel and transverse mixture laws. Thus for cemented carbides which consist of hard carbide particles in a continuous phase of a much softer metallic phase the composite hardness is given by [51, 52]

$$H_c = H_p G V_p + H_m V_m (1 - G) \qquad (4.41)$$

where H_p and H_m are the *in situ* hardnesses of the two phases and G is the contiguity (degree of contact) of the particulate phase defined as [51]

$$G = 2S_{pp}/2S_{pp} + S_{mp} \qquad (4.42)$$

where S_{pp} is the average area of a particle that is in contact with other particles and S_{mp} is the average area of a particle that is in contact with the matrix. Since G is always a fraction < 1, equation (4.41) always falls below the upper bound, linear rule of mixtures. It is to be noted that the *in situ* hardnesses will be influenced by the grain size and/or interparticle spacing. It is assumed that (4.41) also applies to ceramic particulate composites.

4.4.1.2 Fracture resistance measurement. Indentation often results in the formation of relatively large cracks outside the indentation contact area. A common example is the formation of radial cracks growing perpendicular to the surface, for example those formed at the corners of the Vickers hardness impression (Figure 4.16). Such cracks grow both during loading and unloading of the indentor. It is observed that the crack length developed increases with the indentation load. If the indentation crack resistance could be related to the bulk fracture toughness the indentation test would be a very convenient method of fracture toughness measurement and numerous analytical and semi-empirical expressions have been derived (see e.g. the reviews [50, 53, 54]), most having the general form

$$K_{Ic} = \text{const.} \; E^n H^m (P/L)^q \tag{4.43}$$

where n, m and q are constants varying between the different models. It is found in most cases that the equations are not universal, that is cannot be translated between different families of materials [55]. Often the predicted q-value is inappropriate for a given material, that is the derived K_{Ic} value is not independent of indentation load. Nevertheless, the indentation fracture toughness test can yield relative values of toughness, for example in screening a series of composites based on similar constituents or for studying anisotropy, and is frequently used for these purposes.

A second form of indentation is Hertzian indentation with a spherical indentor or flat cylindrical punch [53, 56, 57] which at a critical load produces a ring crack outside the contact area. Since the indentation is purely elastic it can be analysed more rigorously than an indentation involving complex damage. The fracture energy can be determined

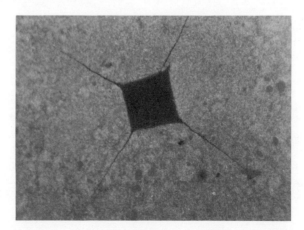

Figure 4.16 Radial crack formed at the corner of a Vickers hardness impression in a SiC_w/Al_2O_3 composite (micrograph courtesy of T. Hansson, Chalmers University of Technology).

either from the critical load [56] or from the crack growth/load dependence [58]. However, both methods are experimentally arduous.

4.4.1.3 Controlled defects and pre-cracks. Indentation is a convenient method of producing controlled flaws for the study of fracture processes and well-defined pre-cracks for fracture toughness tests [59]. Most commonly a notched bar bend test is used. The 'notch' can be simply a hardness indentation crack or this can be developed further into a full straight-through crack by means of so-called bridge-indentation [60]. The advantage of the latter is that it can be treated with stress intensity factors established for standard tests. Moreover, for composites it ensures that the pre-crack is large in relation to the microstructural features.

Another benefit of indentation cracks is that they are convenient for the study of crack paths and profiles which is particularly useful in ceramic composites [61] (see Figure 4.8).

4.4.2 *Compression*

Ceramics are generally considered to have compressive strengths considerably higher than their tensile strengths. This is not necessarily so for long-fibre composites. Loading at an angle slightly away from the fibre direction can lead to interfacial shear failure. Similarly, transverse loading can lead to premature matrix failure as has been shown by Lankford [62]. The effect can be explained in terms of elastic contact damage caused by the fibre when this has a higher elastic modulus than the matrix.

4.5 Friction, wear and erosion

4.5.1 *General nature of wear and erosion*

Good wear resistance is one of the key properties expected of ceramics. The term wear however encompasses a large number of configurations and several mechanisms sensitive to a range of parameters. Here only an abbreviated account is presented to provide a guide to appreciating the wear and erosion behaviour of composites. More detailed accounts are found for example in [63, 64].

Important examples of wear configurations, shown schematically in Figure 4.17, are:

- *Reciprocating* or cyclic wear in which one of the wear surfaces is drawn repeatedly over a confined region of the other as, for

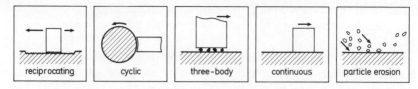

Figure 4.17 Examples of different wear configurations.

example in a pin-on-disc test or a bearing. A special example is fretting in which the extent of motion is very limited.

- *Three-body wear* in which abrasive (or lubricating) particles are caught between two wear surfaces.
- *Continuous, single-pass wear* in which one wear surface is exposed continuously to a previously unworn surface of the other material. This situation arises for example in metal cutting or wire drawing.
- *Particle erosion* in which a surface is worn by the impact of many impinging abrasive particles.

It should also be noted that in these wear situations the wear surfaces can be of the same material or different materials with different properties. Of particular relevance is the relative hardness of the materials.

Some important wear mechanisms that have been identified are illustrated schematically in Figure 4.18 and are described briefly below.

4.5.1.1 Sliding wear. Sliding wear occurs on a surface B that is not penetrated mechanically by asperities on the other, A; that is B is of similar or higher hardness. During sliding, wear can occur for example by (a) microscopic adhesion or welding followed by detachment; (b) chemical reaction and/or interdiffusion, or (c) microfracture and fatigue at or just below the surface. The last of these can be expected in ceramics if conditions of elastic indentation fracture are created. Chemical reaction and interdiffusion occur mainly at high temperatures due either to the ambient conditions or to generation of heat locally (e.g. in metal cutting). High temperature can also be expected to increase adhesive-type wear.

4.5.1.2 Abrasive wear. Abrasive or ploughing wear occurs when asperities on one of the surfaces (or debris between the surfaces) penetrate a softer counter surface leading to pushing out of material. This can occur either plastically, when material will be pushed aside or simply cut out, or by brittle microfracture. The material removed will then accumulate as debris between the surfaces. Two flat surfaces

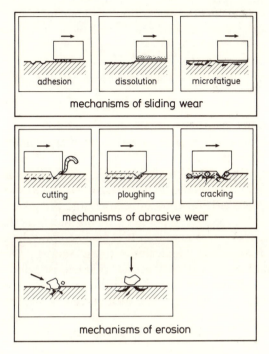

Figure 4.18 Examples of wear mechanisms.

moving against each other can generate abrasive wear since even apparently smooth surfaces are made up of asperities.

Zum Gahr [63] summarises several analytical models for abrasive wear. All are necessarily complex but those appropriate to ceramics share in common an inverse relationship between wear rate and both hardness and fracture toughness of the material being worn; thus;

$$W \propto P/K_{Ic}^{n} \, H^{m} \qquad (4.44)$$

where W is the volume of material loss per unit area and unit time and P is the external contact pressure. Experimental results are in general consistent with this and parameters involving K_{Ic} and H are frequently used as figures of merit for wear resistance. Since the models refer to abrasive wear it is probably more appropriate to use the *difference* in hardness of the two materials rather than H.

4.5.1.3 Particle erosion. Particle erosion of ceramics occurs mainly by microcracking under the impact of the impinging particles. For a given particle size and velocity the degree of damage increases with the angle of impingement up to 90° with an approximately $\sin^2 \alpha$ dependence. At low angles of impingement a proportion of ploughing may occur in some cases. Models of erosion generally yield equations of the form [65]:

$$W \propto V^a r^b \rho^c / K_{Ic}^m H^n \tag{4.45}$$

where V, r and ρ are the velocity radius and density of the impinging particles.

4.5.1.4 Friction. The friction between two surfaces as well as being of practical significance *per se* also provides an indirect indication of wear mechanisms. Perhaps the most common friction measurement is between a pin or stylus moving over a surface of the same material, that is in transition between sliding and abrasive wear. Results typical for ceramics in vacuum as a function of temperature are given schematically in Figure 4.19. At room temperature in air most ceramics exhibit a coefficient of friction of around 0.2. It is assumed that this is determined by a layer of adsorbed molecules on the surface. In vacuum the coefficient increases to values that are characteristic for each ceramic and lie between about 0.4 and 0.7. With increasing temperature the friction first decreases slightly and then increases significantly around about 1000 °C. The reason for the initial decrease is not clear but the increase at 1000 °C is probably due to an increase in adhesion as a result of interdiffusion (i.e. sintering).

4.5.1.5 Wear in metal cutting An important application of ceramic composites is their use as cutting tool materials; it is therefore appropriate to consider the wear conditions specific to metal cutting. Figure 4.20a illustrates the configuration of metal cutting while Figure 4.20b shows the various regions of wear on the cutting tool tip. The metal chip

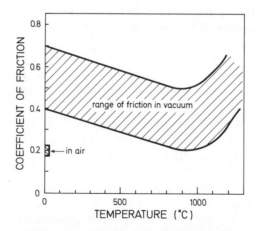

Figure 4.19 Coefficient of friction of ceramics.

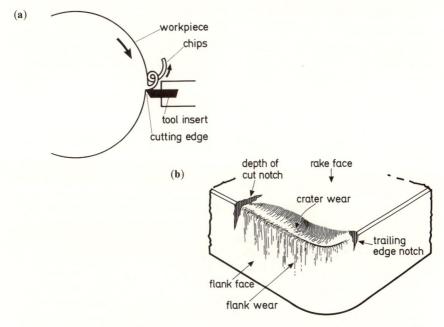

Figure 4.20   (a) The configuration of metal cutting; (b) wear regions on a cutting tool.

passes continuously over the edge of the rake face where it creates a region of crater wear. At the highest cutting speeds this region can attain temperatures up to about 1500 °C [66]. Here probable wear mechanisms are (i) abrasion due to hard phases in the workpiece and possibly particles of the tool material; (ii) sliding wear mechanisms (adhesion, dissolution and chemical reaction). In addition, because of the extremes of temperature, microchipping occurs as a result of thermal fatigue. The relative contribution of each mechanism varies with cutting conditions.

Depth-of-cut and trailing-edge notch wear occur where the two edges of the chip move over the edge of the tool. Here abrasive and sliding wear mechanisms are probably assisted by exposure to the environment (air and cutting fluids, etc.).

The flank face of the tool is worn by the rotating workpiece. Conditions are similar to those on the rake face though the temperature is usually lower.

It should be noted that the geometry of the cutting tip changes as the wear proceeds. This can have a significant effect on the cutting conditions and therefore on the wear mechanisms themselves. In general, wear resistance is favoured by good thermal conductivity in the tool material since this lowers the temperature level and thermal gradients.

4.5.2   *The wear of composites*

The complexity of wear means that few reliable general predictions regarding the wear of composites can be made. The scaling of abrasive wear resistance with hardness and toughness expressed in equation (4.46) cannot be translated directly to composites. The effect of hardness can be explained in qualitative terms by reference to Figure 4.21. Consider a monolithic material, B, being worn by a monolithic material A. If the hardness of A, H_A is less than that of B, H_B, then abrasive wear is negligible and wear occurs predominantly by sliding wear mechanisms. When the ratio H_A/H_B passes unity there is a relatively sharp transition to abrasive wear.

The situation changes if A and/or B are composites since the composite will normally contain a constituent that is harder than the average hardness of the composite. Abrasive wear will begin when the hardest constituent in A exceeds the softest in B which will occur at a ratio of the average hardnesses $H_A/H_B < 1$.

A similar situation exists with respect to the influence of fracture toughness. It was seen in earlier sections that composite toughening often applies only to defects and cracks that cover several microstructural features whereas the microfracture processes in wear often occur in the individual constituents which have lower intrinsic toughness. Moreover, a *reduction* in wear resistance compared to the unreinforced matrix may obtain in cases where the toughening relies on weak interfaces between the constituents or on microcracking.

An implication of the above discussion is that the wear and erosion of composites is usually dominated by the behaviour of the least wear-resis-

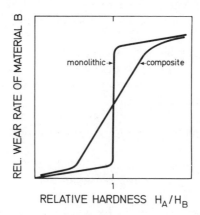

Figure 4.21   Abrasive wear of a material B as a function of the relative hardness H_A/H_B of the abrading material A, comparing the cases of material B being monolithic and composite, respectively.

tant phase. This in turn implies that with respect to volume fraction the wear resistance will generally follow a rule-of-mixtures below the upper bound, linear law, that is it will lie between the upper and lower bounds:

$$\frac{(1/W_1)V_1 + (1/W_2)V_2 > (1/W_c) > (1/W_1)(1/W_2)}{V_1(1/W_2) + V_2(1/W_1)} \qquad (4.46)$$

where here the wear resistance is expressed as the inverse of the wear rate (weight loss per unit area and unit time). The position and shape of the curve can only be predicted through a detailed knowledge of the various wear mechanisms involved bearing in mind that the *in situ* mechanisms in the different constituents will generally be interactive.

The effectiveness of wear resistance reinforcement in a composite is generally favoured by good interfacial strength since this reduces micro-fracture and pull-out. Similarly, in fibre composites the wear resistance at a given volume fraction increases with fibre aspect ratio and with the angle subtended by the fibre axis to the wear surface. For fibres lying in the plane of the wear surface the wear resistance is least for fibres lying at right angles to the direction of motion of the surfaces [63].

It should be noted that deviation from the conditions expressed by equation (4.46), either positive or negative are possible. For example, excessive pulling out of a hard reinforcement as a result of a weak interface could lead to wear resistance below the lower bound while if the second phase acts as a lubricant, wear resistance above the upper bound could be experienced.

A few specific examples of the wear of ceramic composites are given in the chapters of this book but hitherto few examples have been reported in the literature.

4.6 Thermal stability and oxidation

Provided the constituent phases of a ceramic composite are chemically compatible then the limit of thermal stability of the material will be set either by the melting point or decomposition of the constituents or by reaction with the surrounding environment, usually oxidation. Melting points, decomposition temperatures and vapour pressures for most potential constituents of composites are to be found in compilations of thermochemical data and phase diagrams and will not be discussed here. For the majority of composites considered in this book these temperatures are well above 1500 °C.

Prediction of the *oxidation and high-temperature corrosion behaviour* of composites is less straightforward. In general the oxidation behaviour of one constituent phase will be influenced by the other phases both thermodynamically and kinetically. The interfacial boundaries between the phases and impurities also affect the oxidation significantly in many cases. Thus the oxidation of a composite cannot normally be deduced from the behaviour of the constituents.

To simplify the discussion it is convenient to divide the constituent phases of ceramic composites into three classes with respect to oxidation:

- *Oxides*, which do not oxidise; however, in the presence of other oxides, or of other impurities together with oxygen, mixed oxides and glasses with low melting points can form.
- *Non-oxides of Si*, in particular SiC, Si_3N_4 and $MoSi_2$; provided the oxygen partial pressure in the system is not too low, these form an effective protective layer of SiO_2 which limits the rate of oxidation.
- *Other non-oxides*, have a relatively poor oxidation resistance, that is significant rates of oxidation at or below 1000 °C. Examples of relevance to ceramic composites include TiC, TiN, B_4C, BN, TiB_2 etc.

The oxidation behaviour of SiC and Si_3N_4 has been reviewed recently by Vaughn and Maahs [67]. The oxidation is characterised by a transition from passive to active oxidation marked by a decomposition and evaporation of SiO_2:

$$2SiO_2 \rightarrow SiO \uparrow + O_2 \qquad (4.47)$$

The transition is determined by the temperature and the partial pressure of oxygen as indicated in Figure 4.22. In the passive region the

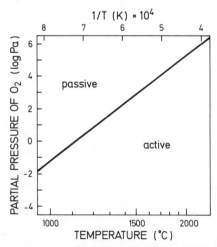

Figure 4.22 The transition between the active and passive oxidation of SiC and Si_3N_4 as influenced by temperature and oxygen partial pressure (after [67]).

oxidation rate is low corresponding to a layer growth rate around 10^{-12}–10^{-11} $(g/cm^2)^2/s$ in 1 atmosphere of air between 1000 and 1500 °C. However, the oxidation in this region is sensitive to the nature of the oxide layer. The rate is much lower if the layer is crystalline rather than amorphous. For layers in the amorphous state, the rate of oxidation increases significantly with decreasing viscosity brought about by certain glass-forming species.

The oxidation resistance of SiC and Si_3N_4 is usually worsened in their composites with oxides. This is because the SiO_2 layer often reacts with the oxide constituent to form a glass or a mixed oxide. Even if the final reaction product is crystalline, intermediate low viscosity, glassy phases may form with a consequent increase in oxidation rate. The situation may be worsened if such a glassy phase penetrates grain boundaries and interfaces.

The effects can be illustrated by reference to Al_2O_3/SiC composites which have been studied in some detail [68, 69]. The oxidation of these composites becomes significant in air at about 1200 °C. Following the initial formation of SiO_2 this begins to react with the alumina to form mullite ($3Al_2O_3$, $2SiO_2$) but also an intermediate non-equilibrium glassy phase (aluminosilicate). In some composites this phase has also been observed to be enriched in Ca. The glassy phase permits rapid diffusion of oxygen to the underlying SiC but also accelerates oxidation by penetrating interfaces and grain boundaries thus providing paths for oxygen into the material. Thus SiC particles or whiskers are rapidly consumed and converted finally to mullite:

$$3Al_2O_3 + 2SiC + 6O \rightarrow Al_6Si_2O_{13} + 2CO \qquad (4.48)$$

This process requires diffusion of oxygen into and CO gas out of the material through the reaction product scale. It appears that both processes occur quite readily through the glassy phase and through cracks in the mullite. The oxidation rate is about an order of magnitude faster than for pure SiC. The final reaction product consists of mullite with an excess of Al_2O_3 or SiO_2 depending on the original proportion of SiC in the composite. The rate of oxidation in SiC/oxide composites is not necessarily accelerated in this way in all cases. In a composite of SiC_w in a mixed Al_2O_3–ZrO_2 matrix the SiO_2 layer crystallised and provided an effective diffusion barrier in the range 1000–1200 °C [70].

Composites containing proportions of non-oxide compounds with poorer oxidation resistance can exhibit satisfactory oxidation resistance provided the non-oxide constituent exists as isolated particles or fibres and provided the oxide it forms does not interact unfavourably with the matrix oxide. In such cases, the oxidation rate decreases once the oxidation of the compound at the free surface is complete.

A further important aspect of oxidation specific to composites is the possible effect of oxidation of the interfaces between the constituent phases. Thus in certain composites, oxidation can be expected to alter the strength of the interface. The discussions earlier in this chapter predict that this could have a strong influence on the subsequent mechanical behaviour of the composite. An important example of this is given by composites reinforced with polymer-precursor SiC fibres which rely for their pseudo-plastic behaviour on a weak graphitic interface. During oxidation this can be rapidly replaced by a strong oxide interface resulting in embrittlement (see chapters 7 and 8).

4.7 Physical properties

This section presents a brief account of those physical properties of ceramic composites that are relevant to their use as structural materials.

4.7.1 Linearly additive properties

Certain properties of composites such as density and specific heat that are not dependent on transport phenomena or on mechanical interactions between the constituents are dependent only on the amounts of the constituents and are given by simple additive mixture laws:

$$\rho_c = \rho_1 V_1 + \rho_2 V_2 \qquad (4.49)$$

$$C_p = C_{p1} V_1 + C_{p2} V_2 \qquad (4.50)$$

4.7.2 Electrical resistance/conductivity

4.7.2.1 Continuous, parallel fibre composites. In the direction of the fibre axis the electrical resistance is given by the simple parallel resistance law:

$$1/\omega_c = V_f/\omega_f + V_m/\omega_m \qquad (4.51)$$

that is, the resistance is dominated by the phase with the lowest resistivity. If one of the phases is electrically conducting then the composite will also conduct in this direction.

In the transverse direction, if there is no continuous connection between the fibres (see below), the series resistance law will apply approximately:

$$\omega_c = V_f \omega_f + V_m \omega_m \qquad (4.52)$$

If the matrix is an insulator then the composite will also be an insulator. If both phases are conductors the contribution of the fibres will be sensitive to the fibre/matrix interface; it can be significantly reduced if the interface gives poor electric contact.

4.7.2.2 Particulate and short-fibre composites. The electrical properties of particulate composites are sensitive to the particle morphology; above all to the degree of connectivity of conducting particles. Predictive models have been reviewed recently by McLachlan *et al.* [71]. Consider a microstructure of isolated conducting particles in an insulating matrix; then the electrical resistance is given approximately by equation [4.52], that is the composite is an insulator. Above a certain volume fraction the particles (or short fibres) will form continuous chains through the material and the composite will become a conductor with a resistivity given approximately by equation (4.51).

Laws of geometrical probability dictate that the transition from isolated particles or groups of particles to continuous chains will occur sharply over a relatively narrow range of volume fraction, the transition being termed the percolation threshold. The critical volume fraction usually lies between 0.1 and 0.5 and is dependent on the particle shape (e.g. it is generally lower for fibres than particles) and the tendency to form shared boundaries (which in turn depends on the relative values of the energies of the particle/matrix interface and the interparticle boundaries [72]). In a composite with a conducting particle morphology, the conductivity is sensitive to the state of the interparticle boundaries, the conductivities of which are sensitive to impurities [47, 73].

4.7.3 Thermal conductivity

The thermal conductivity of ceramic composites has been reviewed recently by Hasselman [47]. Similar principles apply as to electrical conductivity but the differences in conductivity of the different constituents is seldom as dramatic.

Again for the parallel and transverse direction in a unidirectional fibre composite the thermal conductivities are given approximately by the upper and lower mixture law bounds respectively:

$$K_{tc} = K_{tf}V_f + K_{tm}V_m \tag{4.53}$$

$$1/K_{tc} = V_f/K_{tf} + V_m/K_{tm} \tag{4.54}$$

More precise study of the transverse direction shows however that the geometrical arrangement of the fibres can have an influence [74]. It should also be noted that some fibres, notably carbon fibres, exhibit anisotropy of properties. Carbon fibres exhibit a much higher conductivity in the axial direction.

For isolated spherical particles in a matrix with good thermal contact across the interfaces:

$$K_{tc} = \frac{K_{tm}[2K_{tm} + K_{tp} - 2V_p(K_{tm} - K_{tp})]}{2K_{tm} + K_{tp} + V_p(K_{tm} - K_{tp})} \tag{4.55}$$

The thermal conductivity is sensitive to microcracks, porosity and poor interfacial contact in the composite. An indication of this can be obtained by applying equation (4.55) to a material containing spherical pores, that is putting $K_{tp} = 0$:

$$K_{tc} = 2K_{tm}(1 - V_p)/(2 + V_p) \tag{4.56}$$

For convenience, the relationships expressed by equation (4.53)–(4.56) are plotted in Figure 4.23.

The thermal diffusivity, κ, is obtained by definition from

$$\kappa = K_t/\rho C_p \tag{4.57}$$

4.7.4 *Thermal expansion*

For a continuous, uniaxial fibre composite the coefficient of linear thermal expansion in the direction of the fibre axis can be derived using a simple one-dimensional strain model as:

$$\alpha_c(\text{parallel}) = (\alpha_f V_f E_f + \alpha_m V_m E_m)/(E_f V_f + E_m V_m) \tag{4.58}$$

which reveals that the expansion is dominated by the constituent with the higher elastic modulus. The coefficient in the transverse direction is given to a good approximation by the linear mixture law:

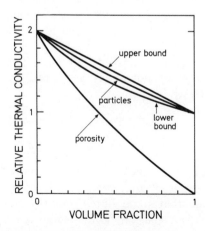

Figure 4.23 Thermal conductivities of composites according to equations (4.53)–(4.56).

$$\alpha_c(\text{transverse}) = \alpha_f V_f + \alpha_m V_m \qquad (4.59)$$

Again, it is to be noted that carbon fibres exhibit a much higher elastic modulus and lower expansion coefficient in the axial direction than in the transverse direction.

An estimate of the linear thermal expansion coefficient of particulate composites assuming that each phase expands or contracts hydrostatically has been given by Kingery [75]:

$$\alpha_c = (\alpha_p B_p V_p + \alpha_m B_m V_m)/3(B_p V_p + B_m V_m) \qquad (4.60)$$

where B_p and B_m are the bulk moduli of the particles and matrix phases, respectively.

References

1. Broek, D. (1986) *Elementary Engineering Fracture Mechanics*, M. Nijhoff, Dordrecht.
2. Weibull, W. (1939) A statistical theory for the strength of materials, *Ingenjörsvetenskapsakad* 151, 1–45.
3. McLean, A. F. and Harstock, D. L. (1989) Design with structural ceramics, in: *Structural Ceramics: Treatise on Materials Science and Technology 29*, ed. Wachtman, J. B., Academic Press, London, pp. 27–97.
4. Warren, R. and Andersson, C.-H. (1984) Silicon carbide fibres and their potential for use in composites, *Composites* 15, 16–24.
5. Aveston, J., Cooper, G. and Kelly, A. (1971) Single and multiple fracture, in: *The Properties of Composites*, IPC Science and Technology Press, Guildford, pp. 15–26.
6. Budiansky, B., Hutchinson, J. W. and Evans, A. G. (1986) *J. Mech. Phys. Solids* 34, 167–189.
7. Corten, H. T. (1967) Micromechanics and fracture behaviour of composites in: *Modern Composite Materials*, eds. Broutman, L. J. and Krock, R. H., Addison-Wesley, Reading, MA, pp. 27–105.
8. Grimvall, G. (1986) *Thermophysical Properties of Materials*, North-Holland, Amsterdam.
9. Christiansen, R. M. (1990) A critical evaluation for a class of micro-mechanics models, *J. Mech. Phys. Solids* 38, 379–404.
10. Takao, Y., Chou, T. W. and Taya, M. (1982) Effective longitudinal Young's modulus of misoriented short fibre composites, *J. Appl. Mech.* 536–540.
11. Sturesson, P. O. and Ståhl, J. E. (1989) Simulation of residual stresses in ceramic composites (in Swedish), report 89–18, Dept. of Production and Materials Engineering, University of Lund, Sweden.
12. Selsing, J. (1961) Internal stresses in ceramics, *J. Am. Ceram. Soc.* 44, 419.
13. Luh, E. Y. and Evans, A. G. (1987) High temperature mechanical properties of a ceramic matrix composite, *J. Am. Ceram. Soc.* 70, 466–469.
14. Korczynski, Y., Harris, S. J. and Morley, J. G. (1981) The influence of reinforcing fibres on the growth of cracks in brittle matrix composites, *J. Mater. Sci.* 16, 1533–1547.
15. Marshall, D. B. and Cox, B. N. (1987) Tensile fracture of brittle matrix composites: influence of fibre strength, *Acta Met.* 35, 2607–2619.
16. Evans, A. G., He, M. Y. and Hutchinson, J. W. (1989) Interface debonding and fibre cracking in brittle matrix composites, *J. Am. Ceram. Soc.* 72, 2300–2303.
17. Thouless, M. D. and Evans, A. G. (1988) Effects of pull-out on the mechanical properties of ceramic matrix composites, *Acta Met.* 36, 517–522.

112 CERAMIC-MATRIX COMPOSITES

18. Lawn, B. R. and Wilshaw, T. R. (1975) *Fracture of Brittle Solids*, Cambridge University Press, Cambridge.
19. Evans, A. G. and McMeeking, R. M. (1986) On the toughening of ceramics by strong reinforcements, *Acta Met.* 34, 2435–2441.
20. Becher, P., Hsueh, C.-H., Angelini, P. and Tiegs, T. N. (1988) Toughening behaviour in whisker-reinforced ceramic matrix composites, *J. Am. Ceram. Soc.* 71, 1050–1061.
21. Faber, K. T. and Evans, A. G. (1983) Crack defection process, part I and II, *Acta Met.* 31, 565–584.
22. Hutchinson, J. W. (1987) Crack tip shielding by microcracking in brittle solids, *Acta Met.* 35, 1605–1619.
23. Taya, M., Hayshi, S., Kobayashi, A. S. and Yoon, H. S. (1990) Toughening of a particulate reinforced ceramic matrix composite by thermal residual stress, *J. Am. Ceram. Soc.* 73, 1382–1391.
24. Sigl, L. S., Mataga, P. A., Dalgliesh, B. J., McMeeking, R. M. and Evans, A. G. (1988) On the toughness of brittle materials reinforced with a ductile phase, *Acta Met.* 36, 945–953.
25. Kriven, W. M. (1986) Displacive transformation mechanisms in zirconia ceramics and other non-metals, in: *Tailoring Multiphase and Composite Ceramics*, eds. Tressler, R. E. *et al.*, Plenum, New York, pp. 223–237.
26. Evans, A. G. (1990) Perspective on the development of high toughness ceramics, *J. Am. Ceram. Soc.* 73, 187–206.
27. Evans, A. G. and Cannon, R. M. (1986) Toughening of brittle solids by martensitic transformations, *Acta Met.* 34, 761–800.
28. Becher, P. F. (1986) Toughening behaviour in ceramics associated with the transformation of tetragonal ZrO_2, *Acta Met.* 34, 1885–1891.
29. Kageyama, K. and Tsu-Wei Chou (1990) Mechanical and statistical analyses of fracture of whisker reinforced ceramic matrix composites, *Int. J. Fracture* 46, 297–315.
30. Giannakopoulos, A. E. and Breder, K. (1991) Synergism of toughening mechanisms in whisker reinforced ceramic matrix composites, *J. Am. Ceram. Soc.* 74, 194–202.
31. Becher, P. F. and Tiegs, T. N. (1987) Toughening behaviour involving multiple mechanisms: whisker reinforcement and zirconia toughening, *J. Am. Ceram. Soc.* 70, 651–654.
32. Charalambides, P. G. and Evans, A. G. (1989) Debonding properties of residually stressed brittle matrix composites, *J. Am. Ceram. Soc.* 72, 746–753.
33. Cook, R. F. and Clarke, D. R. (1988) Fracture stability, R-curves and strength variability, *Acta Met.* 36, 555–562.
34. Bornhauser, A., Kromp, K. and Pabst, R. F. (1985) R-curve evaluation with ceramic materials at elevated temperatures by an energy approach using direct observation and compliance calculation of the crack length, *J. Mater. Sci.* 20, 2586–2596.
35. Cannon, W. R. and Langdon, T. G. (1988) Creep of ceramics, Part 2, an examination of flow mechanisms, *J. Mater. Sci.* 23, 1–20.
36. Ohji, T., Sakai, M., Ito, M., Yamanchi, Y., Kanematsu, W. and Ito, S. (1987) Yielding phenomena of hot pressed Si_3N_4, *High Temp. Technol.* 5, 139–144.
37. Blumenthal, W. and Evans, A. G. (1984) High temperature failure of polycrystalline alumina II, creep crack growth and blunting, *J. Am. Ceram. Soc.* 67, 751–759.
38. Becher, P. F., Angelini, P., Warwick, W. H. and Tiegs, T. N. (1990) Elevated temperature delayed failure of alumina reinforced with 20 vol.% SiC whiskers, *J. Am. Ceram. Soc.* 73, 91–96.
39. Wang, Y. R., Liu, D. S., Majidi, A. P. and Chou, T. W. (1989) Creep characterization of short fibre reinforced ceramic composites, *Ceram. Eng. Sci. Proc.* 10, 1154–1163.
40. de Arellano-López, A. R., Cumbrera, F. L., Rodriguez, A. D., Goretta, K. C. and Routbort, J. L. (1990) Compressive creep of SiC whisker reinforced Al_2O_3, *J. Am. Ceram. Soc.* 73, 1297–1300.
41. Porter, J. R., Lange, F. F. and Chokshi, A. H. (1987) Processing and creep performance of SiC-whisker reinforced Al_2O_3, *Am. Ceram. Soc. Bull.* 66, 343–347.

42. Xia, K. and Langdon, T. G. (1988) The mechanical properties at high temperature of SiC whisker reinforced alumina, *Mater. Res. Soc. Symp. Proc.* 120, 265–270.
43. Lewis, D. (1983) Cyclic mechanical fatigue in ceramic–ceramic composites, *Ceram. Eng. Sci. Proc.* 4, 874–881.
44. Ewart, L. and Suresh, S. (1987) Crack propagation in ceramics under cyclic loads, *J. Mater. Sci.* 22, 1173–1192.
45. Suresh, S. and Brockenbrough, J. R. (1988) Theory and experiments of fracture in cyclic compression: single phase ceramics, transforming ceramics and ceramic composites, *Acta Met.* 36, 1455–1470.
46. Suresh, S. (1990) Mechanics and micromechanisms of fatigue crack growth in brittle solids, *Int. J. Fracture* 42, 41–56.
47. Hasselman, D. P. H. (1986) Tailoring of the thermal transport properties and thermal shock resistance of structural ceramics, in: *Tailoring Multiphase and Composite Ceramics*, eds. Tressler *et al*. Plenum, New York, pp. 731–754.
48. Kerans, R. J., Hay, R. S., Pagano, J. and Parthasarathy, T. A. (1989) The role of the fibre-matrix interface in ceramic composites, *Am. Ceram. Soc. Bull.* 68, 429–442.
49. Doerner, M. F. and Nix, W. O. (1968) A method for interpreting the data from depth sensing indentation instrument, *J. Mater. Res.* 1, 601–609.
50. McColm, I. J. (1990) *Ceramic Hardness*, Plenum, New York.
51. Lee, H. C. and Gurland, J. (1978) Hardness and deformation of cemented tungsten carbide, *Mater. Sci. Eng.* 33, 125–133.
52. Warren, R. (1983) Indentation hardness of microstructures consisting of hard particles in a metallic matrix, in: *Proc. 4th Riso Int. Symp. on Metallurgy and Materials Science*, Riso National Lab., Roskilde, Denmark, 572–582.
53. Cook, R. F. and Pharr, G. M. (1990) Direct observation and analysis of indentation cracking in glasses and ceramics, *J. Am. Ceram. Soc.* 73, 787–817.
54. Ponton, C. B. and Rawlings, R. D. (1989) Vickers indentation fracture toughness test, Parts 1 and 2, *Mater. Sci. Technol.*, 5, 865–872, 961–976.
55. Ekberg, I. L., Lundberg, R., Warren, R. and Carlsson, R. (1989) Indentation testing of SiC-whisker reinforced alumina composites, in: *Brittle Matrix Composites II*, eds. Brandt, A. M. and Marshall, I. H., Elsevier, Barking, 84–97.
56. Warren, R. (1978) Hertzian indentation of brittle solids, *Acta Met.* 26, 1759–1769.
57. Mouginot, R. and Maugis, D. Fracture indentation beneath flat and spherical punches, *J. Mater. Sci.* 20, 4354–4376.
58. Matzke, H. and Warren, R. (1982) Hertzian crack growth in thorium dioxide observed by serial sectioning, *J. Mater. Sci. Lett.* 1, 441–444.
59. Almond, E. A., Roebuck, B. and Gee, M. (1986) Mechanical testing of hard materials, in: *Science of Hard Materials—Institute of Physics Conference Series 75*, Adam Hilger, Bristol, pp. 155–177.
60. Warren, R. and Johannesson, B. (1984) The creation of stable cracks in hardmetals using 'bridge' indentation, *Powder Met.* 27, 25–29.
61. Bengtsson, S., Johannesson, B. and Warren, R. (1986) Profile analysis of fracture surfaces in brittle solids, in: *Structure and Crack Propagation in Brittle Matrix Composite Materials*, eds. Brandt, A. M. and Marshall, I. M., Elsevier Applied Science, London, pp. 69–80.
62. Lankford, J. (1987) Temperature, strain rate and fibre orientation effects in the compressive fracture of SiC fibre reinforced glass-matrix composites, *Composites* 18, 145–152.
63. Zum Gahr, K. H. (1987) *Microstructure and Wear of Materials*, Elsevier, Amsterdam.
64. Buckley, D. H. and Miyoshi, K. (1989) Tribological properties of structural ceramics, in: *Structural Ceramics: Treatise on Materials Science and Technology 29*, ed. Wachtman, J. B., Academic Press, London, pp. 293–365.
65. Wiederhorn, S. M. and Hockey, J. (1983) Effect of material parameters on the erosion resistance of brittle materials, *J. Mater. Sci.* 18, 766–780.
66. Brandt, G. (1990) Ceramic cutting tools, in: *Encyclopedia of Materials Science and Engineering, suppl, vol. 2*, ed. Cahn, R. W., Pergamon, Oxford, pp. 761–765.
67. Vaughn, W. L. and Maahs, H. G. (1990) Active to passive transition in the oxidation

of silicon carbide and silicon nitride in air, *J. Am. Ceram. Soc.* 73, 1540–1543.

68. Lin, F., Marieb, T., Morrone, A. and Nutt, S. (1988) Thermal oxidation of Al_2O_3-SiC whisker composites: mechanisms and kinetics, *Mater. Res. Soc. Symp. Proc.* 120, 323–332.
69. Luthra, K. L. and Park, H. D. (1990) Oxidation of silicon carbide reinforced oxide matrix composites at 1375 °C to 1575 °C, *J. Am. Ceram. Soc.* 73, 1014–1023.
70. Baudin, C. and Moya, S. (1990) Oxidation of mullite–zirconia–silicon carbide composites, *J. Am. Ceram. Soc.* 73, 1417–1420.
71. McLachlan, D. S., Blaszkiewicz, M. and Newnham, R. E. (1990) Electrical resistivity of composites, *J. Am. Ceram. Soc.* 73, 2187–2203.
72. Warren, R. (1976) Determination of the interfacial energy ratio in two phase systems by measurement of interphase contact, *Metallography* 9, 183–191.
73. Takeda, Y. (1988) Development of high thermal-conductive SiC ceramics, *Am. Ceram. Soc. Bull.* 67, 1961–1963.
74. Grove, S. M. (1990) A model of transverse thermal conductivity in unidirectional fibre reinforced composite, *Composite Sci. Technol.* 38, 199–209.
75. Kingery, W. D. (1960) *Introduction to Ceramics*, Wiley, New York.

5 Structural design with advanced ceramic composites

G. C. ECKOLD

5.1 Introduction

The concept of using particulate or fibrous material to reinforce a ceramic matrix is not novel. Amongst the earliest applications of such systems are the use of natural fibres such as grass or animal hair to improve the strength of pottery prior to firing [1]. Current interest in ceramic composites stems primarily from their potential application in advanced engines and aerospace structures, particularly where the component of concern will be exposed to high temperatures [2].

Monolithic ceramics are well known for their refractory behaviour, but their use for engineering structures is severely limited by their poor thermal shock resistance and low fracture toughness. Despite attempts to improve fracture properties, for example transformation toughening of zirconia [3], they are still relatively brittle compared with other engineering materials with fracture strains typically within the range 0.1–0.2% [4]. Figure 5.1 shows the typical stress–strain behaviour of ceramics compared to that of metals and the effect of a low strain to failure can be clearly seen. Due to the effects of brittle behaviour the concept of strength as a material property no longer has the same meaning as for conventional engineering materials and this poses many problems for the design engineer.

Fibre reinforcement of ceramic matrices offers the potential of tougher materials together with an increased strain at fracture [5]. The main mechanisms by which these improvements occur are those due to matrix microcracking and subsequent load transfer within the material. These energy absorbing mechanisms give rise to a form of pseudo-plasticity and result in non-linear elastic behaviour. Figure 5.2 shows the stress–strain behaviour of a glass ceramic composite containing SiC fibres [6]. During matrix microcracking, there is a non-linear region of reduced modulus associated with extension of fibres. A maximum stress occurs corresponding to fibre failure and this may be followed by fibre 'pull-out'. In terms of design the important characteristics of this behaviour are as follows:

- Comparatively high work of fracture due to load-carrying ability beyond matrix cracking.

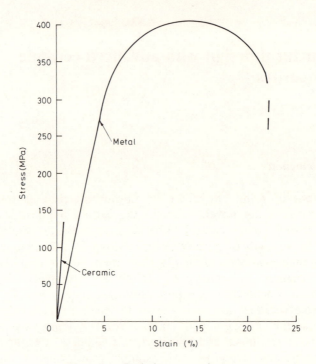

Figure 5.1   Typical stress–strain curves.

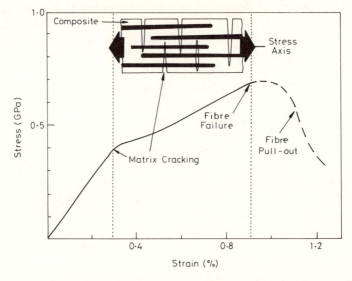

Figure 5.2   Stress–strain behaviour for glass ceramic containing unidirectional SiC fibres
[6].

- Increased failure strain and tolerance of overstressing.
- Reduced sensitivity to flaws within the material.

These effects are advantageous in a number of design situations, not only for simple static load cases, but also in areas of high stress concentration, thermal shock, impact and abrasive wear. Table 5.1 shows typical values of fracture properties [7] for a range of materials. As can be seen significant improvements can be achieved by using reinforcement either in fibre or whisker form. In the latter case mechanical behaviour is broadly similar to monolithic ceramics except with higher toughness and less scatter in strength [8, 9].

Examples of some ceramic-matrix composites currently under consideration include:

(1) Low temperatures (up to 400 °C): carbon fibre reinforced borosilicate glass;
(2) Intermediate temperatures, (up to 1100 °C): silicon carbide reinforced glass ceramics (lithium alumino silicate, cordierite and mullite);
(3) High temperatures (up to 1500 °C): oxidation resistant carbon/carbon (not yet commercially available).

Whilst the benefits of fibre reinforcement are significant it is important that the resulting properties are put into perspective, particularly with regard to those obtained from other engineering materials such as metals and polymer based composites. As can be seen from Table 5.1 fibre reinforcement of ceramics offers significant benefits, but for the

Table 5.1 Toughness properties for a range of materials [7]

Material	Work of fracture ($kJ\ m^{-2}$)	Fracture toughness ($MPa\ m^{1/2}$)
Mild steel	200	200
Al alloy	150	50
Wood	6	7.7
Glass	0.01	0.3
Ceramic (typical)	0.1	4.0
Carbon fibre/ borosilicate glass	5	
SiC fibre/ borosilicate glass	18–26	50
SiC whisker/Al_2O_3	–	9.5
SiC whisker/Si_3N_4	–	6.5
Al_2O_3 fibre/glass	1	7.0
GRP[a]	150	
CFRP[a]	80	

[a] Parallel to fibre direction.

purposes of design they must still be considered as comparatively brittle materials.

Recent interest in ceramic composites is directed towards application in aerospace and high temperature engine components. Typical examples of these include:

- Reinforcement of flame sprayed ceramic coatings using ceramic fibres or wire meshes for supersonic aircraft applications [10]. The perceived advantages of a reinforced coating are increased resistance to spallation and improved thermal shock performance.
- Toughened ceramic components for a number of internal components for military jet engines [11].
- A whole range of under bonnet automotive components [12, 13], for example pistons, valves, turbochargers, where high temperature and wear resistance coupled with good strength are important.
- High precision machine tools [14] where a near zero thermal expansivity would be attractive for new technology applications.
- Reusable, single stage, orbital vehicles, for example HOTOL [15], where fuselage and wing structures, leading edges and nose cones will experience high temperatures on re-entry.
- Process plant equipment [16, 17] where high temperature capability is important.

A feature of these application areas is the importance of component design and of all the materials listed in Table 5.1, ceramic-matrix composites are perhaps the least developed in this respect. In the succeeding sections the use of design methods for ceramic composites will be discussed with special reference to those currently in use for other materials, notably monolithic ceramics and polymer-matrix composites.

5.2 Design methods

In all design assessments there are a number of stages which must be undertaken in order to derive a satisfactory structure both in terms of performance and cost. Those areas which should be considered include:

- functional specification
- materials of construction
- design and analysis
- fabrication limitations
- reliability requirements
- cost considerations

Inevitably there will be a number of options available to the engineer which will result in the design being an iterative procedure. Where a metal is the chosen material of construction it is likely that, due to the extensive knowledge of material properties, analysis methods and fabrication techniques that are available, the number of stages in the design will be reduced to a minimum. In a ceramic composite, however, this will not be the case and, with the level of technology as it now stands, great care must be taken with each of the stages listed above.

Functional specification: the first step in the design of any component is to define clearly the requirements of the application. For composites it is rarely satisfactory to declare that the component must simply perform as a direct replacement for an existing structure, say manufactured from a metal, as this may result in a design which is not the best achievable. Particular attention should not only be given to the magnitude of the primary mechanical and thermal loads, but also to the direction of the bad path. Loads are usually combined and secondary loading such as through thickness components of stress, impact or thermal transients should also be considered.

Materials of construction: for metals there is a wealth of information concerning behaviour under a variety of conditions, much of which is available in formal documentation such as standards and codes of practice. This greatly eases the material selection procedure. With composites, particularly ceramic-matrix composites, there is a general lack of property data and hence it is often necessary to embark on a materials testing exercise in conjunction with the design. Another important consideration with composites is the ability to tailor material properties, for example by changing fibre orientation or type of reinforcement, thereby allowing the material itself to be designed to suit operating conditions.

Design and analysis: the two most pronounced features of the behaviour of composites are anisotropy and heterogeneity. These affect elastic, strength and thermal properties and must be considered in design. This is borne out by experience with polymer composites where it can be shown that failure may occur in the directions other than that of the applied load, for example a Poisson ratio or through thickness effect [18].

Fabrication limitations: It is well known that the method of fabrication and the processing conditions employed can have a marked effect on composite properties [1]. Also, fabrication routes for ceramic-matrix composites are at an early stage of development and not proven in a production environment. Clearly the availability of a suitable production route for the artefact of concern must be an important feature of the design.

Reliability requirements: the essence of any design exercise is the assessment of reliability and the potential risk of component failure.

This is most important as the engineer must take these into account when assessing the factors of safety which should be applied to the design. To a certain extent the selection of factors of safety is a matter of judgement based on experience, and with novel material systems such as ceramic composites this can pose difficulties. The relative brittleness of these materials can also have ramifications on reliability related calculations.

Cost considerations: at the end of the design process the resulting component must be competitive both in terms of performance and cost. Structures fabricated from composites can be expensive compared with those of traditional materials, particularly on the basis of initial procurement cost, and should be viewed in terms of total life costs and improved performance.

There are a number of approaches which may be used for design [19]; a brief summary of these are given in the remainder of this section.

5.2.1   *Empirical design*

Empirical design is a trial and error technique that employs an iterative approach and entails a fabrication and testing programme. A minimum of technical analysis is employed. Although this approach can be satisfactory when supplied property data for an analysis are not available, it is rarely possible to simulate the full range of operating conditions in the laboratory. A consequence of assumptions on scaling, superposition or extrapolation of results can be the imposition of high safety factors on the design.

5.2.2   *Deterministic design*

In this approach the stress distribution within the component is calculated using analytical or numerical techniques. A failure criterion, for example Von Mises, maximum stress or Tresca [20], is normally used in conjunction with experimentally determined strength data. From this a margin of safety can be determined and a judgement made on the acceptability of a candidate material. This technique is often used for metals where the relatively low scatter in property values means that the material can be employed close to its ultimate strength. Indeed, factors of safety as low as 1.5 are not uncommon. For brittle materials, however, the wide distribution in measured strength values can lead to high factors of safety if a low risk of failure is to be assured. Another factor which cannot be taken into account with deterministic design is the dependence of strength on geometry and scale, a feature of brittle materials.

5.2.3 *Probabilistic design*

For brittle materials where strength measurements exhibit wide variability and a marked dependence on specimen size a statistical approach
is often employed to assist design. The most commonly used method is
Weibull analysis [21]. From this it can be shown that the probability of
failure at a particular stress is given by

$$P_f = 1 - \exp\left[- \int_v \frac{(\sigma - \sigma_u)^m}{\sigma_0}\, dv\right] \tag{5.1}$$

where P_f is the probability of failure, σ_u is the threshold stress (lowest
value of stress for which the failure probability is finite), σ_0 is the
normalising parameter (dimensions of stress), m is the Weibull modulus
(this is associated with the variability of failure stresses) and v represents volume. Equation (5.1) can be rewritten in a more useful form and
in terms of physically significant parameters as follows [22]:

$$P_f = 1 - \exp\left[- \int_v \left(\frac{1}{m!}\right)^m (\sigma/\bar{\sigma}_f)^m\, dv\right] \tag{5.2}$$

where $\bar{\sigma}_f$ is the mean value of failure stress and $(1/m)!$ is the gamma
function of $((1/m) + 1)$. Figure 5.3 shows equation (5.2) as a function of
different values of Weibull modulus. It provides a useful guide to the
variability of strength values, the lower the value of m the greater the
variation.

When applying this method to structures, further information is
required concerning the stress distribution. The level of stress in each
element within the component needs to be compared with material
strength data and the probability of failure determined. These values are
then integrated over the structure (stress volume integral) to derive a
probability of failure for the component as a whole.

The main advantage of a probabilistic method is that it allows areas of
high localised stress to be treated more pragmatically. For example, a
material with a low Weibull modulus could still be considered acceptable
for a component with a stress concentration, because the volume of the
element of concern is such that it would not provide a major contribution to the total probability of failure.

As well as advantages there are also a number of problems with this
design technique which should be considered:

- The determination of stress distributions within components must
  be done with a high degree of accuracy. With metals the inherent
  ductility of the materials, which allows some local redistribution of
  stress, means that the need for accuracy is not as severe.

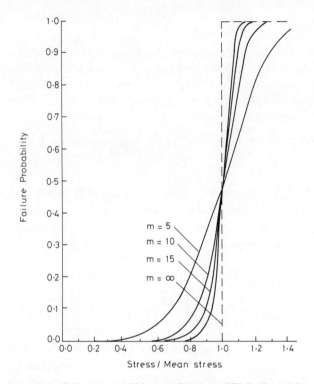

Figure 5.3   Failure probabilities as a function of Weibull modulus.

- A large number of specimens (> 20) must be tested to determine the Weibull modulus. Large uncertainties in the value of m employed in the calculation can negate the value of the analysis.
- The Weibull relationship assumes a uniform, random flaw distribution. Should the population of flaws take a different form the use of a single modulus value may not be appropriate.

The use of probabilistic design methods for ceramic-matrix composites, together with examples, is described in section 5.5.

5.2.4   *Fracture mechanics*

The use of fracture mechanics is a widely employed technique where critical defects within the material are considered in the assessment of structural integrity. For any particular section of a component, defects of various sizes will be present and from a knowledge of applied stress distribution, the stress intensity factor or strain release rate at each flaw can be determined. By examining the propagation of defects during service an assessment can be made of the end of life flaw sizes and these can then be compared to critical values. Fracture mechanics methods have been used successfully in the design of a range of components, for example rotors and pressure vessels [23], where it has been used in conjunction with NDT methods for crack detection.

Use of these methods for ceramics in order to gain an understanding of fracture phenomena is well established [24, 25]. A particularly interesting application in terms of design is in the development of strength/probability/time (SPT) diagrams [13, 26]. This technique attempts to relate time of failure, failure load and the probability of failure. For example, consider a material which when loaded by a given stress breaks in a short time. A similar section of the same material when stressed to say 0.5 of the original load would still fracture, but only when subjected to load for a considerably longer period of time. This time dependence of strength is clearly important in design and is due to sub-critical crack growth occurring under stress.

Derivation of a SPT diagram requires a combination of data relating to fracture mechanics properties and the statistical variation of strength. An example of the end result of the analysis is shown in Figure 5.4, in this case for alumina specimens in bending. The survival probability of a component is plotted as a function of applied stress and time. To demonstrate its application consider a requirement of 95% survival probability and a component lifetime of 10^4 s. From Figure 5.4 it can be seen that the permissible working stress for this system is approximately 205 MPa as compared to a mean short term strength of about 380 MPa.

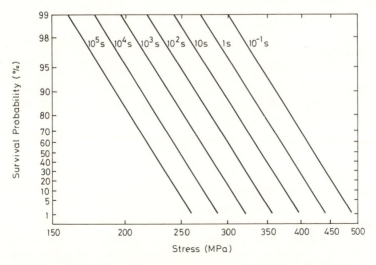

Figure 5.4   SPT Diagram for 95% alumina in bending [24].

5.3 Property determination

The properties of composites are strongly dependent on many parameters such as fibre volume fraction, fibre length, packing arrangement
and fibre orientation. For high performance applications the reinforcement usually consists of continuous aligned or woven fibres in discrete
layers. To design effectively with these systems it is essential that the
material can be characterised in terms of its thermal and mechanical
properties.

As will be discussed later this needs to be carried out both parallel
and transverse to the fibre direction and the ability to calculate these
properties from those of reinforcement and matrix is clearly of importance. Much of the work on this aspect of design follows closely that
carried out for polymer based composites [18, 27, 28].

5.3.1 *Elastic properties of unidirectional composites*

Assuming strain compatibility between fibre and matrix [18], the modulus in the direction of the reinforcement is given to a first approximation by the rule of mixtures expression:

$$E_1 = E_f v_f + E_m(1 - v_f) \tag{5.3}$$

where E_1 is the modulus of the composite in the direction of the fibre,
E_f and E_m are the moduli of fibre and matrix respectively, and v_f is the
fibre volume fraction.

Figure 5.5 shows a comparison between experimentally determined
modulus values [29] and those obtained from equation (5.3). The
correlation between theory and experiment is good. The deviation which
is present has been attributed to the affects of porosity and fibre
misalignment.

As discussed above, the determination of elastic properties in other
directions is also important; assuming continuity of stress between fibre
and matrix [18], expressions can be derived for the transverse and shear
modulus values:

$$E_2 = \frac{E_f E_m}{E_m v_f + E_f(1 - v_f)} \tag{5.4}$$

$$G_{12} = \frac{G_f G_m}{G_m v_f + G_f(1 - v_f)} \tag{5.5}$$

where E_2 and G_{12} are the transverse and shear modulii of the
composite, and G_f and G_m are the shear moduli of fibre and matrix.

It should be noted that the boundary conditions used in the derivation
of equations (5.4) and (5.5) are approximations, and as a result values

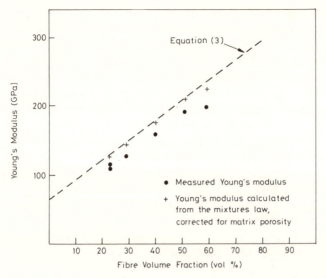

Figure 5.5   Modulus as a function of volume fraction (carbon fibre/glass) [27].

obtained must be used with care. Generally, the experience with polymer-matrix composites is that these expressions underestimate values obtained in practice [28, 30]. There have been numerous attempts to improve the predictive models for these properties, but it is often necessary to resort to semi-empirical methods for accurate results.

The major Poisson ratio can also be determined using a similar approach:

$$v_{12} = v_f v_f + v_m(1 - v_f) \qquad (5.6)$$

where v_{12} is the major Poisson ratio of the composite, and v_f and v_m are those values of fibre and matrix, respectively.

To obtain the minor Poisson ratio the following elastic identity may be used:

$$v_{21} = \frac{E_2}{E_1} v_{12} \qquad (5.7)$$

where v_{21} is the minor Possion ratio of the composite.

Ceramic-matrix composites are often considered for high temperature applications and therefore thermal properties are also important. Again using the rule of mixtures approach yields for thermal expansion:

$$\alpha_1 = \frac{E_f \alpha_f v_f + E_m \alpha_m(1 - v_f)}{E_f v_f + E_m(1 - v_f)} \qquad (5.8)$$

$$\alpha_2 = \alpha_f v_f + \alpha_m(1 - v_f) + v_f \alpha_f v_f + v_m \alpha_m(1 - v_f)$$
$$- [v_f v_f + v_m(1 - v_f)]\alpha_1 \qquad (5.9)$$

where α_1 and α_2 are the longitudinal and transverse thermal expansivities of the composite, and α_f and α_m are those for fibre and matrix, respectively.

A number of models have been proposed to predict the thermal conductivity of composites as a function of the conductivities and morphologies of the constituents. Some of these are presented in chapter 4. Since conductivity can be very sensitive to other factors such as porosity and the nature of the interface, the predictions of the models should be employed with caution.

All of the equations presented above relate to properties in the primary material directions, that is parallel and perpendicular to the fibre direction (Figure 5.6). In design it is often necessary to compute values in other directions, and this can be accomplished using conventional transformation relationships.

For those properties which are second order tensors such as thermal expansion the following expression may be used:

$$\begin{bmatrix} \alpha_x \\ \alpha_y \\ \alpha_{xy} \end{bmatrix} = [T] \begin{bmatrix} \alpha_1 \\ \alpha_2 \\ 0 \end{bmatrix} \tag{5.10}$$

where $[T]$ is the transformation matrix for a second order tensor. In expanded form it may be written as

$$[T] = \begin{bmatrix} m^2 & n^2 & 2mn \\ n^2 & m^2 & -2mn \\ -mn & mn & m^2 - n^2 \end{bmatrix} \tag{5.11}$$

where $m = \cos\theta$ and $n = \sin\theta$.

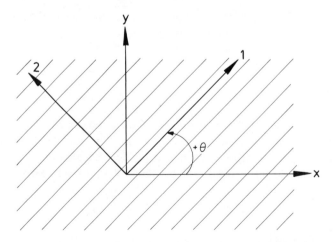

Figure 5.6 Relationship between primary material directions (1, 2) and off-axis directions (x, y).

For fourth order tensors, for example stiffness and compliance, the relationships are more complicated. In matrix form the expressions are as follows:

(a) For stiffness:

$$[\bar{Q}] = [T]^{-1}[Q][T]^{-T} \tag{5.12}$$

(b) For compliance:

$$[\bar{S}] = [T]^{T}[S][T] \tag{5.13}$$

where $[Q]$ and $[S]$ are stiffness and compliance matrices whose components relate to the primary material directions (1,2), and $[\bar{Q}]$ and $[\bar{S}]$ are those for the off-axis direction (x, y). $[T]$ is the transformation matrix (equation (5.11)).

For plane stress, the stress–strain equations may be written in expanded form as

$$\begin{bmatrix} \sigma_1 \\ \sigma_2 \\ \sigma_{12} \end{bmatrix} = \begin{bmatrix} Q_{11} & Q_{12} & 0 \\ Q_{12} & Q_{22} & 0 \\ 0 & 0 & Q_{66} \end{bmatrix} \begin{bmatrix} \varepsilon_1 \\ \varepsilon_2 \\ \varepsilon_{12} \end{bmatrix} \tag{5.14}$$

$$\begin{bmatrix} \sigma_x \\ \sigma_y \\ \sigma_{xy} \end{bmatrix} = \begin{bmatrix} \bar{Q}_{11} & \bar{Q}_{12} & \bar{Q}_{16} \\ \bar{Q}_{12} & \bar{Q}_{22} & \bar{Q}_{26} \\ \bar{Q}_{16} & \bar{Q}_{26} & \bar{Q}_{66} \end{bmatrix} \begin{bmatrix} \varepsilon_x \\ \varepsilon_y \\ \varepsilon_{xy} \end{bmatrix} \tag{5.15}$$

Coefficients within the stiffness matrix may be related to engineering constants as follows:

$$Q_{11} = \frac{E_1}{1 - v_{12}v_{21}}, \qquad Q_{12} = v_{21}Q_{11}$$

$$Q_{22} = \frac{E_2}{1 - v_{12}v_{21}}, \qquad Q_{66} = G_{12} \tag{5.16}$$

In the matrices containing off-axis properties the presence of terms such as an apparent shear expansitivity, α_{xy}, and stiffnesses linking shear strains and normal loads, $\bar{Q}_{16}$, $\bar{Q}_{26}$, should be noted. The significance of these constants, which are zero for isotropic materials or when properties on the primary material axes are being considered, is discussed in section 5.4.1.

5.3.2 Elastic properties of laminated composites

In section 5.3.1 the means by which the elastic properties of a unidirectional composite ply can be determined were discussed. However, in practice it is common for plies of different angle to be used in the form of a multidirectional laminate, the number and disposition of plies being dictated by the applied loading conditions. For such a laminate the full stress–strain relationship may be written as

$$\begin{bmatrix} N \\ M \end{bmatrix} = \begin{bmatrix} A & B \\ B & D \end{bmatrix} \begin{bmatrix} \varepsilon \\ \kappa \end{bmatrix} \qquad (5.17)$$

where $[N]$ and $[M]$ are forces and moments per unit width defined below, and $[\varepsilon]$ and $[\kappa]$ are extensional strains and curvatures, respectively.

$$N = \int \sigma\, dz, \qquad M = \int \sigma z\, dz \qquad (5.18)$$

The laminate stiffness matrices can be calculated as follows:

$$[A] = \sum [\bar{Q}]_k (z_k - z_{k-1})$$

$$[B] = \frac{1}{2} \sum [\bar{Q}]_k (z_k^2 - z_{k-1}^2) \qquad (5.19)$$

$$[D] = \frac{1}{3} \sum [\bar{Q}]_k (z_k^3 - z_{k-1}^3)$$

where the summation is carried out through the laminate thickness, $k = 1, \ldots, n$.

Figure 5.7 shows the geometry of a typical laminate construction to which the above expressions would apply.

In equation (5.10), $[A]$ are extensional stiffnesses, $[B]$ are coupling stiffnesses and $[D]$ are bending stiffnesses; the value of z is defined by reference to Figure 5.7. As with the property matrices for single plies, $[\bar{Q}]$, these matrices are symmetric with nine coefficients. Although these appear complex, the matrices simplify considerably if the laminate

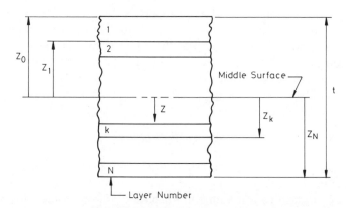

Figure 5.7 Geometry of laminate.

possesses symmetry. For example, a symmetrical cross-ply laminate of unit thickness with equal numbers of 0° and 90° layers has elastic constants defined by

$$A_{11} = A_{22} = \tfrac{1}{2}(Q_{11} + Q_{22}), \qquad A_{12} = Q_{12}$$
$$A_{16} = Q_{16}, \qquad A_{16} = A_{26} = 0$$

(5.20)

In this case all coefficients in the coupling matrix $[B]$ are zero. The significance of laminate asymmetry on non-zero terms in $[B]$ are discussed in section 5.4.1. The use of these expressions is well established for polymer composites [13] and there are numerous microcomputer programs which may be obtained to carry out the calculations [31, 32].

5.3.3 Strength properties

In the previous sections the elastic properties of ceramic composites were considered and relationships given indicating how composite properties could be calculated from a knowledge of those for fibre and matrix. Essentially the expressions listed are those originally derived for polymer based systems, but are equally applicable to materials with a ceramic matrix. With strength properties, however, there are distinct differences in behaviour between the two types of material. A significant factor which must be considered is the comparatively low strain to failure of ceramic matrices. Figure 5.8 shows stress–strain curves for epoxy and glass matrices reinforced with silicon carbide fibres [33]. With a tensile strain of approximately 0.3%,the slope of the curve for the glass composite decreases significantly, whereas that for the epoxy

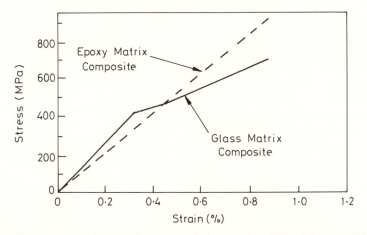

Figure 5.8   Stress–strain curves for epoxy and glass matrices (SiC reinforcement) [31].

system is essentially linear to failure. As discussed earlier, this behaviour is due to multiple matrix microcracking in the ceramic matrix prior to fibre failure.

Using the simple strain compatibility model used to derive equation (5.3) for modulus, the stress at which matrix cracking occurs is given by

$$(\sigma_c^*)_{mF} = \sigma_{mF}[1 + v_f(E_f/E_m - 1)] \tag{5.21}$$

where $(\sigma_c^*)_{mF}$ is the composite stress at which the matrix will crack and σ_{mF} is the ultimate strength of the unreinforced matrix.

The ultimate strength of the composite can also be calculated by a rule of mixtures approach.

$$\sigma_{cF} = v_f \sigma_{fF} \tag{5.22}$$

where σ_{cF} is the ultimate strength of the composite and σ_{fF} that for the fibre.

In terms of design, the ramifications of matrix cracking requires careful thought as, although there is still considerable strength and toughness retention, problems may arise due to propagation of cracks through repeated loading, and the possibility of degradation of exposed fibres through environmental attack or thermal fatigue loading.

Experimentally it is noted that equation (5.21) generally underestimates the matrix cracking stress. In unreinforced ceramics the propagation of a crack from a single flaw may be sufficient to cause total failure, whereas in composites the presence of fibres can inhibit crack growth. The theory describing the development of cracking in fibre composites (Aveston, Cooper, Kelly–ACK) [34, 35] is relatively well established and uses energy considerations to derive expressions from which the matrix cracking strain may be calculated:

$$(\varepsilon_c^*)_{mF} = \left[12 \frac{\tau \gamma_m v_f^2 E_f}{r E_c E_m^2 (1 - v_f)}\right]^{1/3} \tag{5.23}$$

where $(\varepsilon_c^*)_{mF}$ is the failure strain of the matrix, τ the shear strength of the matrix, γ_m the fracture surface energy of the matrix and r the fibre radius.

As can be seen from equation (5.23) the matrix failure strain can be enhanced by increasing fibre volume fraction and reducing fibre diameter. Examination of the situation in terms of a fracture mechanics approach [36, 37] leads to a similar expression for matrix cracking stress:

$$(\sigma_c^*)_{mF} = \frac{12\tau\gamma_m}{r} \left[\frac{v_f^2}{(1 - v_f)} \frac{E_f}{E_c E_m^2} (1 - v^2)^2\right]^{1/3} \frac{E_c}{1 - v^2} \tag{5.24}$$

The results from these expressions give good agreement with experimental results [38].

The above discussion relates to the determination of strengths in the direction of the fibre. Currently, there are no reliable methods for determining strengths in other primary directions, and experimentally determined values should be used [39]. As a first approximation the matrix strength can be used to provide a value transverse to the fibre, but it should be noted that experience from polymer-matrix materials is that this would provide an overestimate.

To obtain strength values in other directions one of the failure criteria for anisotropic materials may be used. These theories, the stress distribution in a material, for example calculated using the transformation matrix (5.11), together with the primary material strengths may be used to predict failure.

Of the many theories available, the two most often used are:

(a) Maximum stress:

$$\frac{\sigma_1}{\sigma_L} < 1, \qquad \frac{\sigma_2}{\sigma_T} < 1, \qquad \frac{\sigma_{12}}{\sigma_S} < 1 \tag{5.25}$$

where σ_L, σ_T and σ_S are the design strengths in the primary material directions.

(b) Tsai-Hill:

$$\left(\frac{\sigma_1}{\sigma_L}\right)^2 + \left(\frac{\sigma_2}{\sigma_T}\right)^2 - \frac{\sigma_1 \sigma_2}{\sigma_L^2} + \left(\frac{\sigma_{12}}{\sigma_S}\right)^2 < 1 \tag{5.26}$$

Both of these criteria have been successfully correlated to experimental results [40, 41]. Figure 5.9 shows the results using the Tsai-Hill criterion for a silicon carbide/glass composite.

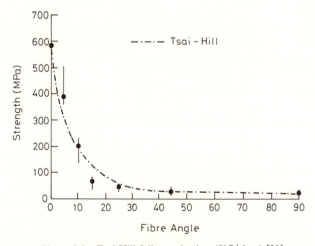

Figure 5.9   Tsai-Hill failure criterion (SiC/glass) [39].

For laminate constructions comprising a series of layers of different angles, the failure behaviour can become quite complex with significant load redistribution occurring as individual lamina fail [42–44]. However, this can be accommodated using the above failure criteria in conjunction with the elastic analysis described by equations (5.11)–(5.19).

5.4 Aspects of component design

In addition to the determination of material properties and their application in laminate calculations there are a number of other facets of behaviour which could be important for a given component. In the following sections some of these aspects of performance are discussed, highlighting their significance in a design situation.

5.4.1 *Anisotropy*

Due to the directional nature of the reinforcement in many ceramic composites, material properties can be highly anisotropic. For the prediction of elastic performance, equations (5.10)–(5.19) are quite satisfactory. However, there are some aspects of behaviour which are not immediately apparent from the analysis and merit special attention [18, 45]:

- *Laminate symmetry:* as already cited in section 5.3.2, for a laminate which is not symmetrical about its mid-plane coupling may occur between normal forces and curvatures, and bending moments and extensional strains. An example of this is a laminate constructed of two plies, $+\theta$ and $-\theta$, each ply being of equal thickness. In this case the application of a tensile load to a flat plate will cause the specimen to twist. Similarly, if the plate is heated it will warp due to the effect of a coefficient of thermal shear, α_{xy}. Generally, laminates of symmetrical lay-up should be used wherever possible.
- *Through thickness effects:* in the design of isotropic materials, it is common for through thickness stresses to be considered as second order effects due to their low magnitude. With composites they may still be low, but the value of through thickness strength can also be small. Therefore care must be taken when designing areas such as geometrical discontinuities, free edges, openings and attachments where through thickness tensile and interlaminar shear stresses are at their peak.
- *Direction of loading:* Composites with continuous reinforcement offer the potential for a high degree of design optimisation as the

fibres can be disposed in precise directions. Implicit in this statement is the assumption that the engineer knows, with a high degree of accuracy, the nature and direction of the applied loads. A compromise in design may be required to cater for any uncertainty. Indeed, it is not uncommon for engineers to resort to quasi-isotropic constructions where loading regimes are complex.

Whilst the effect of anistropic elasticity certainly complicates procedures, the majority of the problems are either amenable to solution or can be avoided by careful design.

5.4.2 Residual stress

In addition to the loading applied during component operation, consideration must also be given to any regime of residual stress which may be present within the material. For ceramic-matrix composites the mismatch in thermal expansivities between fibre and matrix is of key importance in this respect due to the high temperatures at which these materials are processed [1, 46]. Residual stresses of large magnitude can also be generated where the matrix contains more than one phase, for example cristobalite $(\alpha \simeq 27 \times 10^{-6} \, °\mathrm{C}^{-1})$ in borosilicate glass $(\alpha \simeq 3.3 \times 10^{-6} \, °\mathrm{C}^{-1})$.

If the difference in expansivity between matrix and fibre is positive, cooling from the processing temperature causes the matrix to be placed in compression which can create a significant increase in matrix cracking stress on loading. On the other hand the fibres become loaded in tension which may result in lower ultimate strengths. Conversely if the difference in expansivity is negative the matrix is placed in tension and cracking may occur. Radial effects may also be apparent whereby the fibre shrinks away from the fibre reducing the bond strength of the fibre/matrix interface. The matrix stress induced by thermal expansion mismatch may be estimated by [1]

$$(\sigma_t)_m = \frac{(\alpha_m - \alpha_f)E_f v_f \Delta T}{1 + v_f(E_f/E_m - 1)} \tag{5.27}$$

where $(\sigma_t)_m$ is the thermal mismatch stress in the matrix and ΔT is the temperature change.

This stress should then be considered to be additive to that determined from equation (5.24).

In addition to micromechanical effects the use of laminates may give rise to macromechanical thermal stress [18]. This is due to the fact that the individual plies comprising a laminate may have different rates of thermal expansion. The forces and moments generated by this effect may be quantified as follows:

$$[N_t] = \int [\bar{Q}][\alpha]\Delta T\, dz$$

$$[M_t] = \int [\bar{Q}][\alpha]\Delta Tz\, dz$$

(5.28)

These forces and moments may then be treated as per mechanical loads, equations (5.17)–(5.19), and the thermal stresses calculated.

5.4.3 *Combined loads*

In design situations it is unusual for single loads to be applied in isolation; combined, multi-axial loading is more often the norm. For isotropic materials these can be catered for by calculating the values of principal stress and using a suitable failure criterion. Composites can be treated in a similar fashion, although it is more common to base judgements of failure on stresses acting in the primary material directions. The failure criteria given by equations (5.25) and (5.26) may be used for this purpose. Often the result of this analysis is a failure envelope (or failure surface for a three-dimensional stress system) (Figure 5.10). Structures which are subjected to load systems within the

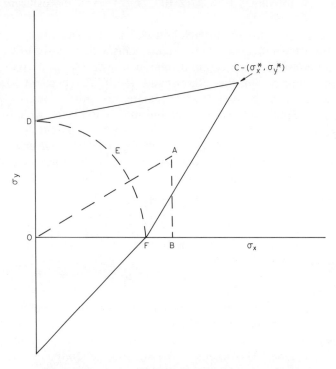

Figure 5.10   Approximate biaxial failure envelope [43].

envelope are considered to be satisfactory whilst those with loads outside the envelope would be deemed to be beyond their design limit. It should be noted that for composites, the manner in which the loads are applied is also important. For example, consider point A, following the load path denoted OA would be satisfactory, but following the route OBA would prove to be unsatisfactory as it would take the structure outside the envelope as the load is applied.

In certain laminate configurations it is possible to define an approximate envelope without recourse to sophisticated analysis [44]. Angle ply laminates where the individual layers are disposed $+\theta$, $-\theta$, $+\theta$, ... are examples where a simplified approach may be adopted. In this case it can be shown that the extreme coordinates of the envelope in Figure 5.10, point C, may be calculated as follows:

$$\sigma_x = \sigma_{cF} \cos^2 \theta, \qquad \sigma_y = \sigma_{cF} \sin^2 \theta \qquad (5.29)$$

The ultimate composite strength, σ_{cF} is used in equation (5.29) as an example; in some cases the matrix cracking stress, $(\sigma_c^*)_{mF}$, may be more appropriate.

This point is of considerable significance in design as it represents that combination of loads which acts in the fibre direction. As a corollary, the optimum angle for a given combination of loads (applied simultaneously) is given by

$$\frac{\sigma_y}{\sigma_x} = \tan^2 \theta \qquad (5.30)$$

The subject of combined loading in polymer composites has received much attention but a definitive solution to the problem is not yet available. However, provided judgement is exercised the approaches outlined above can serve the needs of most design situations.

5.4.4 Jointing

It will almost invariably be the case that ceramic composite components will be employed as part of an assembly comprising a number of different materials. In these circumstances jointing and load transfer within a joint will be of prime importance. Joints may take a number of forms depending on operational requirements; demountable/permanent, load carrying/locational, and between similar/dissimilar materials. Generally speaking a poor design will result if jointing configurations for ceramics are simply copied from those established for metals. This is especially the case for demountable connections where stress concentrations due to bolt holes or rivets would be unacceptable in a ceramic based material.

In polymer composites, the problem of joining to other materials using adhesives has now reached the stage where joints can be designed and manufactured with confidence [47]. Continuity of load transfer is a particularly attractive feature of this method of bonding.

Due to temperature limitations, conventional adhesives are not appropriate for ceramic materials and alternative methods are under development. Examples of these include [13, 48]:

- welding, where there are matched thermal expansivities (applicable in only a limited number of cases);
- active metal brazing using reactive metals which allow wetting of the ceramic;
- diffusion bonding using soft metal interlayers.

To date, much of this research has been carried out using unreinforced ceramic materials and further work is required to demonstrate effectiveness with composites working in a highly stressed environment.

5.4.5 *Impact*

A key feature of performance in many applications for advanced materials is behaviour under impact conditions. Indeed, improvement in impact behaviour is one of the prime reasons for introducing reinforcements into ceramics in the first instance. The ability of a material to withstand impact can be characterised in the laboratory by the measurement of toughness and stress intensity factors. However, in a design situation it is recognised that these are of limited value and there is no single parameter which may be used in the selection of a suitable composite material. Instead a pragmatic approach is often adopted, depending on the application being considered. One approach which is widely accepted is the concept of damage tolerance, defined as the residual strength of the material after an impact event.

A further consideration is the effect of simultaneous tensile loading and impact, a design condition which often occurs in practice, for example a turbine blade under centrifugal stress. It has been shown that this is a much more severe condition than a sequential measurement of impact followed by tensile load [49]. Figure 5.11 shows the results of testing carbon reinforced glass composites of nominally identical laminate configuration. The curves shown in Figure 5.11 represent the boundary between fracture and non-fracture for each particular condition of impact energy and applied stress. An interesting feature of this data is that at low applied stress the borosilicate system is superior to the LAS glass ceramic system, whereas at high applied stress the reverse

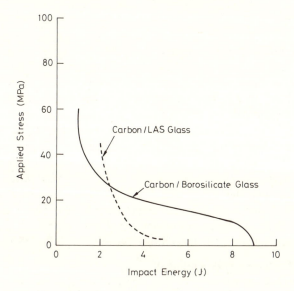

Figure 5.11   Fracture/no fracture boundary for carbon/glass ceramics [47].

is true. The reasons for this are not fully understood, but the ramifications on design are significant. If material selection were to be based on simple, unstressed impact tests the borosilicate system would appear superior, but should the component be subjected to significant stress in service this result would be incorrect.

5.4.6   *Environmental effects*

Retention of properties at high temperatures is the major motivating factor behind the development of ceramic-matrix composites and clearly the effect of temperature must be catered for in design. Generally, all of the design considerations described in the preceding sections are applicable at elevated temperatures providing the basic materials data are available. Measurement of mechanical properties at temperature has received much attention and considerable effort is being made to modify materials to meet the rigours of high thermal loading [50–53]. Thermal shock resistance is one area where the tailoring of constituent materials has proved effective [53, 54]. The presence of microcracks, caused either during processing or as a result of exceeding the matrix cracking strain in service, can be particularly deleterious in adverse environments. Although short term ambient strengths may be retained there are concerns that paths for environmental attack on the fibres may be created.

5.4.7   *Compressive design*

One of the techniques which may be employed to overcome problems
associated with the inherent brittleness of ceramics is to use the material
only under compressive or near zero tensile stress conditions. The
rational behind this is that compressive loads can normally be trans-
mitted perpendicular to the axis of a defect and, due to friction,
significant load can also be carried parallel to a defect. In the majority
of structures, however, an entirely compressive regime of stress is
unlikely to be present and a novel design approach must be adopted. A
technique which is often considered is to suppress tensile stress by
applying a compressive pre-strain. Normally this is achieved by using the
ceramic with other materials and applying pre-load either mechanically
or through thermal expansivity mismatch. Examples where this has been
used with some success for monolithic ceramics include [13]:

- liners for four-stroke engines;
- turbine blades where a hollow ceramic aerofoil is connected to the
  metal hub by a tensioned tie bar (Figure 5.12);
- turbine wheels in which the ceramic hub and blades are kept under
  compression using shrinkage rings;
- multilayer piston/cylinder designs for diesel engines.

Clearly, if this approach is to be adopted due account must be taken of
any system of residual stresses which may be generated in the component during manufacture and the effect of contact reactions with other

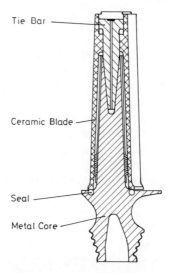

Tie Bar

Ceramic Blade

Seal

Metal Core

Figure 5.12   Tie rod blade design with hollow ceramic aerofoil [11].

materials. It is probable that the final component design will be significantly different from that of an existing metallic counterpart.

It must be noted that with composites, behaviour under compressive loading must be given careful attention. For unidirectional materials low values of compressive strength may be obtained due to matrix microfracture and fibre buckling [55]. Similar effects are observed with polymer matrix systems [18]. A further consideration is the effect of fatigue where recent work [56–59] has shown that undesirable levels of crack growth can occur under cyclic compressive loads significantly less than the static compressive strength.

5.4.8 *Discontinuous reinforcement*

Much of the preceding discussion has been focused on the application of composites with long, effectively continuous fibre reinforcement. This is appropriate due to the potential benefits in performance which may be achieved with these systems. The use of discontinuous reinforcement such as particulates, platelets and most fibres, can also enhance certain properties and therefore may be used to produce useful materials. Several such materials exhibit isotropic or near-isotropic behaviour and, as has already been indicated, design procedures used for monolithic ceramics are often adequate [8, 9].

The main benefit of these systems is a substantial increase in fracture toughness over the unreinforced matrix (chapters 6 and 9). Strengths however, are usually lower than the matrix due to a combination of thermal stress and stress concentrations around fibres. It is only in a small number of cases where these effects can be so minimised that benefits in terms of strengths can be realised [60, 61]. If some alignment of fibres can be achieved during processing a degree of strength enhancement is possible.

Although effectively isotropic, failure of these systems under combined loads can deviate from that observed for other materials. It can be shown using a probabilistic argument that the failure envelope for brittle materials under a biaxial tension load can be approximated to a circular arc (line DEF, Figure 5.10) [62, 63]. This is an important result in design as it suggests a weakening of the material with respect to its uniaxial tensile strength.

5.5 Design examples

In the preceding sections expressions for the calculation of elastic and strength properties for laminates are described. With regard to component design, further work must be carried out to determine the stress distribution within the structure. For components of simple geometry it is likely that this can be accomplished analytically, but where this is not possible numerical techniques such as finite elements are available.

Once the detailed stresses are available they need to be interpreted with respect to design allowables. Of the methods which are available to carry out this task (section 5.2), those based on probabilistic techniques are receiving the most attention. Numerous studies have been carried out on polymer composites and design exercises have been undertaken for components with applied centrifugal, pressure and thermal loads [64–68]. Inevitably the process of design will be an interactive one, the complexity of which will depend on the component and its application.

5.5.1 Beam

The effects of probabilistic design can be most apparent with structures under bending. This is due to the fact that only a relatively small volume of material is subjected to high tensile stress.

For a beam subjected to 3-point bending it can be shown that the relationship between tensile and bending strengths is as follows:

$$\frac{(\sigma_{cF})_B}{(\sigma_{cF})_T} = [2(m + 1)^2]^{1/m} \qquad (5.31)$$

where $(\sigma_{cF})_B$ and $(\sigma_{cF})_T$ are the composites strengths in bend and tension, respectively.

For a value of m of 20 (measured for SiC/glass) this ratio is 1.4, that is the bend strength is 1.4 times the tensile strength. This type of approach has been shown to be satisfactory for monolithic ceramics and is likely to give reasonable results for discontinuous reinforcements [24]. In the case of undirectional composites however, it is found that the bend strength is considerably greater than that predicted from equation (5.31). Recent work [69] has shown that this may be rationalised by considering the composite as a bundle of fibres within the matrix. From this it can be deduced that the stress volume integral of a unidirectional material depends only on the stress distribution along the length of the beam and not through the thickness. This results in a modified expression for bend/tensile strength ratio:

$$\frac{(\sigma_{cF})_B}{(\sigma_{cF})_T} = 2(m + 1)^{1/m} \qquad (5.32)$$

In this case the Weibull modulus relates to that of the fibres and not the composite. Adopting a typical value, 5, the strength ratio becomes 1.6 which gives closer agreement with experiment.

Consideration of bending is of prime importance in structural design as many components can be reduced to a series of beams and panels subjected to such loads.

5.5.2   Pressure vessels

Pressure vessels are ideally suited for manufacture from unidirectional composites as the stress distributions within the structure can be precisely defined [14, 70]. For example, a cylindrical pressure vessel with end closures would have a stress distribution given by

$$\sigma_h = \frac{pd}{2t}, \qquad \sigma_a = \frac{pd}{4t} \tag{5.33}$$

where σ_h and σ_a are hoop and axial stresses, p is the internal pressure, and d and t are the diameter and thickness of the cylinder, respectively.

From equation (5.30) therefore, the optimum disposition of fibres would be $\pm 55°$ from the axis of the cylinder. It should be noted however, that this only applies for the 2:1 stress ratio defined above. Should this change, for example if the vessel is pressurised without end closures, and the axial stress becomes zero, a $\pm 55°$ angle would prove to be a poor design (see Figure 5.10).

Pressure vessels are manufactured in a number of forms each with its own characteristics. For a cylinder where the thickness is comparable to the diameter, equation (5.33) is no longer applicable. For such vessels the stress distribution at radius r is described by

$$\sigma_h = \frac{pr_i^2}{r_o^2 - r_i^2}\left[1 + \left(\frac{r_o}{r}\right)^2\right]$$

$$\sigma_a = \frac{pr_i^2}{r_o^2 - r_i^2} \tag{5.34}$$

$$\sigma_r = \frac{pr_i^2}{r_o^2 - r_i^2}\left[1 - \left(\frac{r_o}{r}\right)^2\right]$$

where r_o and r_i are the outer and inner radius of the cylinder, and σ_r is the radial component of stress.

Using equation (5.2) the probability of failure can be written as

$$P_f = 1 - \exp\left[-\left(\frac{1}{m}!\right)^m (\sigma/\sigma_f)^m \Sigma\right] \tag{5.35}$$

where

$$\Sigma = \int_v \left[\frac{\sigma_h}{\sigma H(\sigma_h)}\right]^m + \left[\frac{\sigma_a}{\sigma H(\sigma_a)}\right]^m + \left[\frac{\sigma_r}{\sigma H(\sigma_r)}\right]^m dV$$

$\sum$ is the stress volume integral and $H(\sigma)$ is a step function dependent on the sign of the stress. For tensile stress $H(\sigma) = 1$; for the compressive stress $H(\sigma)$ equals the ratio of mean failure stress of unit volume in compression to that for tension.

Equation (5.35) can be solved numerically using the appropriate values of geometry and Weibull modulus [22]. Figure 5.13 shows the results of these computations in terms of failure probability as a function of internal pressure.

5.5.3 *Turbine wheel*

Due to their high strength to weight ratio at elevated temperatures ceramic-matrix composites are particularly attractive for gas turbine components. Figure 5.14 shows an outline of one such possible component, a blade from a turbine wheel. In service, the primary load will be centrifugal with the dominant principal stress in the radial direction.

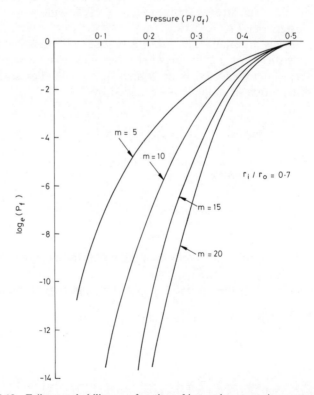

Figure 5.13 Failure probability as a function of internal pressure in a pressure vessel.

However, there will also be some twisting on the blade and so as a first stage in the design, a quasi-isotropic laminate may be selected.

From a full analysis, the stress volume integral may be derived and the probability of failure determined for a given material and rotational speed [71]. The analysis can be taken further to provide useful information on how to modify the structure to improve its performance. It is often the case, particularly when stress distributions are complex, that the stresses in a relatively small region contribute predominantly to the stress volume integral. This is of importance as the percentage contribution to the stress volume integral is approximately equal to the percentage of the total failure initiating from that volume. Figure 5.14 demonstrates this effect by indicating the volumes contributing for different Weibull moduli. A number of important design conclusions may be drawn:

- For high values of m an increased fillet radius would have a significant influence on component strength, whereas for low values of m a more radical change in design would be necessary to have the same effect.
- For high values of m inspection procedures may be concentrated in the fillet area, whereas for lower values of m the whole component would need to receive rigorous examination.

From this type of assessment the next stage of the design may be undertaken. This may not only include the changes in geometry cited above, but also modification of the proposed laminate construction.

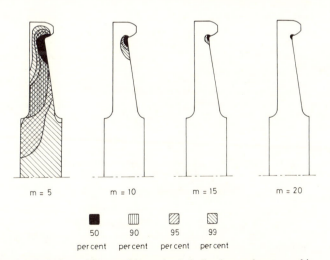

Figure 5.14   Boundaries within turbine wheel indicating volumes making percentage contributions to the total stress volume integral [64].

5.6 Conclusions

Ceramic-matrix composites are currently at the forefront of advanced materials technology. These developments are being stimulated by the requirements of a number of specific high performance applications and as such component design is of key importance. In this chapter the techniques available to the engineer for design are presented with a particular emphasis on those procedures in widespread use for similar materials, notably polymer based composites. Although the applicability of these methods has not been fully demonstrated for ceramic-matrix materials, they do form a firm basis from which the design process can proceed.

References

1. Phillips, D. C. (1983) in: *Fabrication of Composites*, eds. Kelly, A. and Mileiko, S. T., Elsevier, Amsterdam, Chapter 7.
2. Davis, L. W. and Bradstreet, S. W. (1970) in: *Metal and Ceramic Matrix Composites*, Calmers, Boston, Chapter 13.
3. Lankford, J. (1986) in: *Proc. Symp. Advanced Structural Ceramics*, eds. Becher, P. F., Swain, M. V. and Somiya S., Materials Research Society, p. 61.
4. Davidge, R. W. (1987) *Composites* 18, 92.
5. Dawson, D. M., Preston, R. F., Briggs, A. and Davidge, R. W. (1986) AERE Report R12025, Harwell Laboratory.
6. Prewo, K. and Brennan, J. J. (1982) *J. Mater. Sci.* 17, 1201.
7. Bailey, J. E. (1988) *Conf. Ceramic Composites and Coatings*, London.
8. Duffy, S. F., Manderscheid, J. M. and Palko, J. L. (1989) *Ceram. Bull.* 68, 2078.
9. Nemeth, N. N., Manderscheid, J. M. and Gyekenyesi, J. P. (1989) *Ceram. Bull.* 68, 2064.
10. Sklarew, S. (1961) in: *Mechanical Properties of Engineering Ceramics*, eds. Kriegel, W. W. and Palmour, H., Interscience, London, Chapter 13.
11. Marsh, G. (1989) *Aerospace Composites Mater.* 1, 34.
12. Baker, H., ed. (1988) *Guide to Engineered Materials, Advanced Materials and Processes*, 3.
13. Bunk, W. and Hausner, H., eds. (1986) *Ceramic Materials and Components for Engines, Proc 2nd Int. Symp.* Lubeck, DKG, Bad Honnef.
14. Hyde, A. R. (1988) *Conf. Ceramic Composites and Coatings*, London.
15. Walmsley, S. (1988) *2nd Int. Conf. Automated Composites*, Netherlands.
16. Reagan, P., Ross, M. F. and Huffman, F. N. (1988) *Ceram. Eng. Sci. Proc.* 9, 881.
17. DeBelles, C. L. and Kendel, K. E. (1987) *AIChE Symp. Pittsburgh*, p. 256.
18. Jones, R. M. (1975) *Mechanics of Composite Materials*, Scripta, Washington.
19. Richerson, D. W. (1982) *Modern Ceramic Engineering*, Marcel Dekker, New York.
20. Timoshenko, S. (1956) *Strength of Materials*, Van Nostrand, New York.
21. Weibull, W. (1951) *J. Appl. Mech.* 18, 293.
22. Stanley, P., Fessler, H. and Sivill, A. D. (1973) *Proc. Br. Ceram. Soc.* 22, 453.
23. Irvine, W. H. and Quirk, A. (1971) *Conf. Practical Application of Fracture Mechanics to Pressure Vessel Technology*, I Mech E, London.
24. Davidge, R. W. (1979) in: *Mechanical Behaviour of Ceramics*, Cambridge University Press, Cambridge.
25. Atkins, A. G. and Mai, Y. W. (1985) in: *Elastic and Plastic Fracture*, Ellis Harwood, New York.
26. Davidge, R. W., McLaren, J. R. and Tappin, G. (1973) *J. Mater. Sci.* 8, 1699.

27. Calcote, L. R. (1969) in: *Analysis of Laminated Composite Structures*, Van Nostrand, New York.
28. Tsai, S. W. and Hahn, H. T. (1980) *Introduction to Composite Materials*, Technomic, Westport CI, USA.
29. Phillips, D. C. (1972) *J. Mater. Sci.* 7, 1175.
30. Wells, G. M. (1988) PhD Thesis, University of Bath.
31. ESDU, Engineering Sciences Data Unit, ESDU International, London.
32. COMPCAL, Centre for Composite Materials, University of Delaware, USA.
33. Prewo, K. M. (1986) *J. Mater. Sci.* 21, 3590.
34. Aveston, J., Cooper, G. A. and Kelly, A. (1971) Single and multiple fracture, in: *Proc. Conf. on the Properties of Fibre Reinforced Composites*, IPC, Guildford.
35. Hale, D. K. and Kelly, A. (1975) *Annu. Rev. Mater. Sci.* 2, 405.
36. Marshall, D. B., Cox, B. N. and Evans, A. G. (1985) *Acta Metall.* 33, 2013.
37. McCartney, L. N. (1987) *Proc. R. Soc. (London) Ser A* 409, 329.
38. Cooper, G. A. (1971) *Rev. Phys. Technol.* 2, 49.
39. Dawson, D. M., Binstead, A. and Ko, F. K. (1989) *Br. Ceram. Trans. J.* 88, 226.
40. Phillips, D. C., Sambell, R. A. J. and Bowen, D. H. (1972) *J. Mater. Sci.* 7, 1454.
41. Briggs, A. and Davidge, R. W. (1989) *Mater. Sci. Eng.* A109, 363.
42. Prewo, K. M. (1988) *J. Mater. Sci.* 23, 2745.
43. Davidge, R. W. (1988) in: *Application of Fracture Mechanics to Composite Materials*, ed. Friedrich, K., Elsevier, Amsterdam.
44. Sbaizero, O. and Evans, A. G. (1986) *J. Am. Ceram. Soc.* 6, 481.
45. Eckold, G. C. (1985) *Composites*, 16, 41.
46. Goettler, R. W.and Faber, K. T. (1985) *Ceram. Eng. Sci. Proc.* 9, 861.
47. Lee, R. J. and McCarthy, J. C. (1989) in: *Advanced Composites*, ed. Partridge, I. K., Elsevier, Amsterdam.
48. Nicholas, M. G. and Lee, R. J. (1989) *Metals and Materials*, p. 348.
49. Phillips, D. C., Park, N. and Lee, R. J. (1990) *Composites Sci. Technol.* 37, 249.
50. Lankford, J. (1988) *Ceram. Eng. Sci. Proc.* 9, 843.
51. Carroll, D. F., Tressler, R. E., Tsai, Y. and Neir, C. (1986) in: *Tailoring Multiphase and Composite Ceramics*, ed. Tressler, R. E., Material Science Research Series, Vol. 20, p. 775
52. Hillig, W. B. (1986) in: *Tailoring Multiphase and Composite Ceramics*, ed. Tressler, R. E., Materials Science Research Series, Vol. 20, p. 697.
53. Hasselman, D. P. H. (1986) in: *Tailoring Multiphase and Composite Ceramics*, ed. Tressler, R. E., Materials Science Research Series, Vol. 20, p. 731.
54. Donald, I. W. and McMillan, P. W. (1976) *J. Mater. Sci.* 11, 949.
55. Lankford, J. (1987) *Composites* 18, 145.
56. Suresh, S., Ewart, L., Maden, M., Slaughter, W. S. and Nguyen, M. (1982) *J. Mater. Sci.* 22, 1271.
57. Suresh, S. and Brockenbrough, J. R. (1988) *Acta Metall.* 36, 1455.
58. Suresh, S. and Han, L. X. (1988) *J. Am. Ceram. Soc.* 71, 2158.
59. Han, L. X. and Suresh, S. (1989) *J. Am. Ceram. Soc.* 72, 1233.
60. Buljan, S. T. and Sarin, V. K. (1987) *Composites* 18, 99.
61. Phillips, D. C. (1987) *Int. Conf. Composite Materials*, London.
62. Stanley, P. and Sivill, A. D. (1978) *Proc. Br. Ceram. Soc.* 26, 97.
63. Awaji, H. and Sato, S. (1978) *Int. J. Fracture* 14, 13.
64. King, R. L. (1988) *CAD in Composite Material Technology*, Southampton.
65. King, R. L. (1987) *GEC J. Res.* 5, 76.
66. Stanley, P., Sivil, A. D. and Feester, H. (1974) *Proc. 5th Int. Conf. on Exp. Stress Analysis*, UDINE.
67. Margetson, J. and Stanley, P. (1976) *Int. J. Mech. Sci.* 18, 561.
68. Stanley, P., Sivil, A. D. and Tessler, H. (1978) in: *Fracture Mechanics of Ceramics* (3), eds. Bradt, R. C., Hasselman, D. P. and Flange, F., Plenum Press, New York, p. 51.
69. Davidge, R. W. and Briggs, A. (1989) *J. Mater. Sci.* 24, 2815.
70. Soden, P. D. and Eckold, G. C. (1983) in: *Development in GRP Technology*, ed. Harris B., Elsevier Applied Science, London.
71. Stanley, P., Sivill, A. D. and Fessler, H. (1978) *J. Strain Anal.* 13, 103.

6 Particulate ceramic-matrix composites

R. WARREN and V. K. SARIN

6.1 Introduction

This chapter considers composites consisting of ceramic matrices combined with particles, as opposed to fibres and whiskers, of a second
ceramic material. The materials most commonly chosen as matrices are
the main structural ceramics, alumina, silicon carbide and silicon nitride.
Composites with these matrices have been developed and are produced
commercially but numerous experimental studies have been made with
other ceramic matrices such as aluminium nitride and mullite. The
added particulate phase can be chosen from many ceramics and refractory compounds. A limitation is matrix/particle incompatibility, that is
chemical reaction or significant intersolubility at temperatures of processing or service. Nevertheless, the number of compatible combinations is
quite large even if only the most common ceramic and refactory
compounds are considered (e.g. those listed in Table 1.1). The potential
offered by the possibilities of modifying microstructure and the use of
multicomponent composites is almost limitless. Here the discussion will
be limited to the most well known ceramic/ceramic composites; materials reinforced with a ductile metallic phase (e.g. cermets) will not be
considered. Selected examples of composites and relevant properties are
given in Table 6.1 and two typical microstructures are shown in Figures
6.1 and 6.2.

The general principle of composite materials is to enhance selected
properties of the monolithic matrix. In particulate ceramic composites,
benefits that can be achieved include improved toughness without
significant loss of hardness, improved hardness without loss of toughness, and improved thermal conductivity. A combined improvement in
these properties leads to improved wear resistance and improved performance as tool materials (see chapter 4). By appropriate choice of
reinforcement the ceramic can be made electrically conducting thus
providing electrical materials for high temperature, oxidising environments as well as permitting spark machining.

Four possible toughening mechanisms in particulate composites are:

  (i) transformation toughening, the most common example being the
      zirconia toughening of alumina (ZTA);

Table 6.1 Physical and mechanical properties of selected ceramics and particulate ceramic composites (the values are 'likely' values estimated for low porosity material on the basis of data reported in handbooks, suppliers information and reported literature)

Material (compositions in vol%)	Density ρ (g cm^{-3})	E (GPa)	ν	C_p 1000 °C (J/g K)	κ 1000 °C (W/m K)	α 0–1000 °C (K$^{-1} \times 10^{-6}$)	Resistance (Ω m)	Mismatch $\Delta\alpha$ ($\alpha_m - \alpha_p$)	K_{Ic} (MN m$^{-3/2}$)	Hardness (VPN or KPN) (GPa) 20 °C	Hardness 1000 °C	Wear (merit)[b]	Thermal shock (merit)[c]	Ref.[e]
Al$_2$O$_3$	3.99	390	0.23	1.25	6	8.0	>10^{15}	–	3–5	19	7	12	6	–
SiC	3.2	440	0.15	1.25	40	4.5	1	–	3.4	26	–	13	60	–
β-Si$_3$N$_4$	3.18	300	0.22	1.25	15	3	–	–	3.5–6	16	12	10–15	65	–
Al$_2$O$_3$–50B$_4$C	3.28	380	0.2	1.6	8.5	7	90	+2.5	4.5	20	–	14	11	[13, 26]
Al$_2$O$_3$–30TiC	4.26	400	0.22	1.12	9.5	8.5	0.1	−0.5	4.0	21	8	13	9	
Al$_2$O$_3$–24Ti(N$_{0.75}$C$_{0.25}$)	4.3	400	–	1.15	10	8.5	–	−0.5	3.5	19	8	11	7	[28]
Al$_2$O$_3$–3ZrO$_2$	4.0	380	–	1.23	9	8.6	High	–	4.2	18	7	12.5	8	[28]
Al$_2$O$_3$–15ZrO$_2$	4.2	370^a	–	1.15	8	–	High	–	6–10^d	17	–	19	14	[9]
SiC–16TiB$_2$	3.3	430	0.15	1.24^a	40	5	<1^a	−0.5	4.5	27	–	16	70	[30, 35]
SiC–25TiC	3.63	450^a	–	1^a	40^a	5^a	<1^a	−4.0	6	28	–	20	85	[12]
Si$_3$N$_4$–30TiC	3.70	350^a	–	1^a	20^a	5^a	–	−5.5	4.5	17	–	12.5	40	[13–16,39]
Mullite–10SiC	2.84	240	0.27	1^a	–	5^a	–	−1	2.3	15	–	7	–	[25]
Mullite–10ZrO$_2$	3.1	160^a	–	1^a	4^a	5^a	–	–	4	(14)	–	11	16	[25]

^{a}Value derived by rule of mixtures, not experimental.
^{b}Figure of merit = $K_{Ic}^{3/4} \times H^{1/2}$.
^{c}[$(1 - \nu)\kappa \, K_{Ic}/E\alpha$] $\times 10^{-4}$.
^{d}Various particle sizes and degrees of transformation.
^{e}References are examples of studies of the systems and not necessarily the source of data.

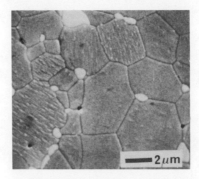

Figure 6.1 Scanning electron micrograph of Al_2O_3–3 vol% ZrO_2; the ZrO_2 is the light phase (thermally etched in H_2).

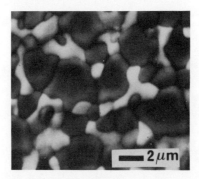

Figure 6.2 Scanning electron micrograph of Al_2O_3–24 vol% Ti(N, C). The carbonitride is the light phase (thermally etched in H_2).

   (ii) microcracking in cases where the thermal expansion mismatch is appropriate;
  (iii) crack deflection;
  (iv) matrix compression due to thermal expansion mismatch $(\alpha_p > \alpha_m)$.

In addition to these direct processes it is important to note that the addition of particles to the ceramic almost always involves changes in the matrix microstructure and thereby causes changes in its properties indirectly. Such changes include, for example grain size and grain size distribution, porosity and its morphology, matrix composition, grain boundary properties, etcetera.

Of the toughening mechanisms, transformation toughening is very effective but is limited to moderate temperatures. The second two mechanisms can give moderate improvements but they have less potential than the same mechanisms and other toughening mechanisms in fibre composites. However, an important advantage of particulate composites over fibre composites is that they can be prepared by the relatively straightforward methods of conventional ceramic processing, that is processing that is more versatile and far less involved and consequently less expensive than methods demanded for long-fibre and whisker reinforced materials. In general, the processing conditions that are required for particulate composites are not very different from those of the unreinforced matrix. The potential health hazard posed by small diameter whiskers is also circumvented in particulate composites.

A development which may in part offer some of the benefits of whisker reinforcement without the need to forfeit convenient processing is the introduction of single crystal platelets as reinforcement (see Figure 6.3). For example the use of Al_2O_3, TiB_2 and SiC platelets in composites has been reported [1–3].

6.2 Processing

6.2.1 *Microstructure*

The properties of particulate composites are dependent not only on the properties and fractions of the constituent phases but also on other features of the microstructure. The microstructure is in turn to a large extent dependent on the details of processing. At the same time mechanisms of densification and microstructural development are often affected by the presence of second phase particles. Thus, as a preface to a description of processing it is appropriate to recall some of the significant microstructural parameters:

- *Porosity and pore size distribution*: porosity has a marked effect on

Figure 6.3   Scanning electron micrograph of SiC platelets (courtesy of American Matrix Inc., Knoxville, TN 37933, USA).

fracture strength, hardness and elastic moduli. When the pores act as strength-determining flaws then the pore size distribution will determine the scatter of strength (e.g. the Weibull modulus). Strength and reliability demands for many structural applications are such that porosities greater than 1–2% are unacceptable and an upper limit must be set on maximum pore size.

- *Matrix grain size*: above a certain size the grains in a ceramic rather than the pores can act as the strength determining flaws. Consequently, a low matrix grain size is generally considered desirable. Rice and Freiman [4] have shown that for non-cubic ceramics a maximum in toughness is attained at intermediate grain sizes (e.g. at aproximately 100 μm for alumina). This effect is due to local, internal stresses caused by the anisotropy in thermal expansion of the grains leading to toughening by microcracking and wake bridging. This leads to *R*-curve behaviour but not necessarily higher strengths or resistance to microfracture phenomena.
- *Matrix grain morphology*: in some ceramics the properties may be influenced by grain morphology. For example the elongated grain shape in some β-silicon nitrides is considered to enhance toughness [5].
- *Reinforcement particle size*: this is of particular importance in zirconia toughened ceramics since the size of the zirconia particles affects the tetrahedral to monoclinic transformation. However reinforcement size also appears to be significant in other systems.
- *Phase continuity*: the existence or absence of continuity (percolation) of conductive particles in a non-conductive matrix is critical for the conductivity of the composite.
- *Properties of grain boundaries and interface*: the composition and properties of interfaces in the microstructure have a large influence on both the mechanical and physical properties of the composite. In their turn the interfaces are influenced by impurities, sintering additives and chemical reactions between the phases.

Such microstructural effects are often sufficient to mask the expected effects of composite mixing as expressed by composite models and rules of mixtures. Thus those composite properties presented in Table 6.1 that were measured experimentally are not necessarily predictable from the properties of the constituent phases.

6.2.2 *Conventional processing techniques*

As discussed in the introduction, an important advantage of particulate composites over whisker and fibre composites is that they can be prepared by powder processing techniques developed for conventional ceramics. In general some modifications are necessary to minimise particle attrition, to ensure thorough mixing of the component particles and to overcome possible inhibiting effects of the particles on pressability and densification. Similarly when choosing conditions of processing, account must be taken of the effects of the particulate phase on phase transformations and other aspects of microstructural development as outlined in section 6.2.1.

A conventional processing route, illustrated by the flow chart in Figure 6.4, may typically involve the following stages (for descriptions of processing of specific systems the reader is referred to the references in Table 6.1; processing and principles are also discussed in chapter 3):

(1) *Matrix milling*: the matrix powder is milled in a suitable inert liquid together with sintering aids where required. The powders are normally submicron with very narrow size distribution and freed of agglomerates by a prior separation process. The sintering aids are usually the same as those used for the monolithic matrix (e.g. 1–10% alumina, yttria and/or magnesia in silicon nitride, around 1% Al and carbon in SiC, very small additions of MgO in alumina).

(2) *Blending*: the matrix powder and reinforcement particles are blended by milling and/or ultrasonic mixing. This stage is critical in determining the uniformity of particle distribution while retaining a desired particle size and morphology. Pressing lubricants and binders can be added at this stage if required.

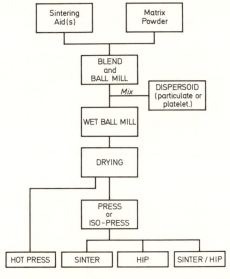

Figure 6.4 Flow chart showing typical processing routes for particulate composites.

(3) *Drying*.
(4) *Pressing to shape*: cold pressing generally yields a green density of 50–55%.
(5) *Consolidation* (e.g. sintering, hot pressing or HIP).

Under favourable circumstances adequate density can be achieved by a cold pressing/pressureless sintering process. For example alumina/zirconia composites can be sintered at around 1600 °C and other alumina based composites at around 1800 °C. Effective densification is favoured by a number of factors such as fine and uniform particle size, homogeneous particle mixing and packing, appropriate sintering atmosphere and sintering aids. The sintering temperatures and times that can be employed are often limited by chemical reactions, intersolubility or decomposition of the constituents. These effects can sometimes be delayed, thereby permitting higher sintering temperatures, by use of moderate overpressure of the sintering atmosphere or of powder beds having a similar composition to the composite. Thus Si_3N_4 can be sintered at around 1650 °C with an overpressure of about 25 atmospheres.

The densification of a ceramic can be expected to be inhibited when adding particles of another ceramic having a lower densification rate. An example is shown in Figure 6.5. Here the densification of TiB_2 (activated by a small addition of Fe) is reduced by additions of B_4C greater than 10 wt%. However, up to 10% of carbide addition enhances densification, a phenomenon observed in a number of systems. Figure 6.5 also shows the marked inhibition of matrix grain growth caused by second phase particles, attributed to the Zener effect and observed for

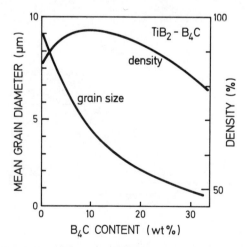

Figure 6.5 Example of the effect of second phase particles on the grain growth and densification of a particulate ceramic composite during pressureless sintering (TiB_2–B_4C, 2000 °C, flowing argon) [6].

almost all matrix/particle combinations. Grain growth inhibition is normally beneficial since it permits greater sintering temperatures and times before unfavourable grain sizes are reached (see chapter 3).

Even in cases where pressureless sintering does not provide adequate densification for the intended application, the density may be sufficient to permit containerless hot isostatic pressing. Such 'sinter-HIP' processes circumvent the need for (uniaxial) hot pressing which is limited to only simple component shapes.

Hot pressing has to be employed to achieve close to full density in materials that cannot be sintered adequately by pressureless sintering. As already indicated, pressureless sintering may be limited by inter-action, decomposition or grain growth of the constituents. The use of simultaneous pressure permits a lower sintering temperature. Typical pressures are 20–200 MPa applied for about 1 h and depending on the sintering aids present and the temperature chosen.

Other processing routes developed for monolithic ceramics such as slip casting, sol-gel processing [7] and injection moulding can also be applied to particulate ceramics. Sol-gel processing has been particularly successful for oxide/oxide composites in which both components can be in the form of colloids. Extremely fine particle sizes and effective densification are achieved. Sol-gel processes are however still relatively expensive. In slip casting some difficulty can be experienced in preparing stable slips of unlike constituents (e.g. oxide/non-oxides) due to differences in surface charge effects. This has recently been overcome in alumina/carbide systems by coating the carbide particles with an alumina-like layer [8].

6.2.3 *Transformation toughened composites*

The processing of transformation toughened ceramics requires special consideration since the subsequent transformation characteristics are affected by processing. For example, tetragonal zirconia particles in a matrix will not normally undergo transformation to the monoclinic state and therefore not contribute to toughening if their diameter is less than about 1 μm. On the other hand if they are too large they will transform spontaneously upon cooling after sintering and again toughening will not be effective. In choosing a sintering temperature and time in combination with a starting particle size to provide an optimum particle size in the sintered composite, consideration must be given simultaneously to the grain growth of the matrix; the matrix grain size must not become large enough to give strength degrading flaws. This has led to the concept of a sintering 'window', framed on two sides by the minimum and maximum permitted zirconia particle size and on the third side by the maximum permitted matrix grain size [9]. In the case of zirconia the size and position of the window can be adjusted by means of the controlled addition of stabilisers such as yttria. Sol-gel techniques produce excellent materials because they give an extremely fine starting particle size, a narrow particle size distribution and a very uniform mixture. The microstructure can be placed in the window without risk that a proportion of the particles will be too large [10].

6.2.4 *New alternative techniques*

A number of novel techniques have been developed that lead to particulate composite microstructures. Important examples are *in situ* reactions, self-propagating synthesis (SPS) [10] and the Lanxide DI-MOX^{TM} process [11] (see also chapter 3).

6.2.4.1 In situ reactions. An example of an *in situ* process is the sintering of a mixture of AlN and B_2O_3 which results in a composite of alumina and boron nitride [10]. The occurrence of the reaction normally assists densification.

6.2.4.2 Self-propagating synthesis. In SPS the composite constituents are produced by an *in situ* reaction that is sufficiently exothermic to maintain a temperature sufficient for densification without external heating. The process does not usually give adequate densification and must be followed by a final densification process. Two examples [10] are:

$$4Al + 3TiO_2 + 3C = 2Al_2O_3 + 3TiC \qquad (6.1)$$

$$10Al + 3TiO_2 + 3B_2O_3 = 5Al_2O_3 + 3TiB_2 \qquad (6.2)$$

6.2.4.3 The Lanxide DIMOX^{TM} process. If a molten aluminium surface is allowed to oxidise the oxide layer will grow through a preform of particles placed on top of the surface. This is the principle of the Lanxide DIMOX^{TM} process. High density composites with alumina matrices can be produced by the method. By suitable choice of metal bath and reacting gas, other composites can be produced, for example Ti and nitrogen will yield TiN-based composites [11].

6.3 Properties

6.3.1 *Room temperature mechanical properties*

Some mechanical properties of selected particulate composites are summarised in Table 6.1 together with corresponding values for the

three main matrices. Comparison with other relevant monolithic materials should be made by reference to Table 1.1. The properties of the
composites are not fully consistent with the rules of mixtures and other
models presented in chapter 4. This is almost certainly due to variations
in microstructure as discussed above even though the data has been
chosen to represent material with as little porosity as possible.

No attempt is made to give results for fracture strength or for strength
scatter. This is because the strength is determined by the fracture
toughness in combination with the defect structure of the material and
the latter is not directly related to the composite composition. The mean
fracture strengths of most high density composites lie within the range
500–1000 MPa with Weibull moduli of 7–15. A detailed study of
SiC–TiC composites [12] has clearly shown that the Weibull modulus is
influenced by details of processing and not the composition.

Particulate reinforcement is an effective means of increasing the
hardness of a ceramic if the added particle has a significantly greater
hardness. A roughly linear rule of mixtures relationship is achieved
between the two constituent hardnesses as is shown for Al_2O_3–TiC and
Si_3N_4–TiC composites in Figure 6.6 [13].

In most composite systems reinforced with non-transformation toughening particles, moderate increases in toughness, up to around 50%, are
observed. Silicon nitride matrix composites appear to be an exception;
significant decreases in toughness with increasing particulate content are
observed (Figure 6.7). However, for this matrix at least it has been

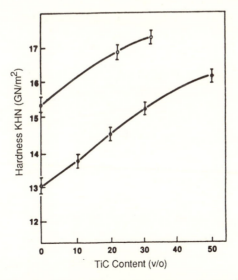

Figure 6.6   Effect of fraction TiC particles on the hardness of alumina and silicon nitride
matrix composites [13]. $\bigcirc$ Al_2O_3 + TiC; $\bullet$ Si_3N_4 + TiC.

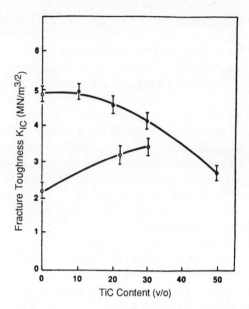

Figure 6.7 Effect of fraction TiC particles on the fracture toughness of alumina and silicon nitride matrix composites [13]. ○ Al$_2$O$_3$ + TiC; ● Si$_3$N$_4$ + TiC.

shown that the fracture toughness is improved by increasing the size of the particles [13–15] (Figure 6.8).

Which mechanism or mechanisms of toughening are operating in these composites is not clear. Most systems exhibit a thermal expansion mismatch with the particle having a larger expansion coefficient which

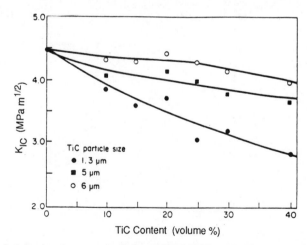

Figure 6.8 Indentation fracture toughness of Si$_3$N$_4$–TiC composites as a function of TiC particle fraction and size [14].

would lead, after sintering, to tensile stress in the particle, compressive hoop stress in the matrix, and a radial tensile stress in the matrix (chapter 4). In these circumstances a propagating crack will tend to be led around particles and therefore crack deflection is a probable toughening mechanism. This is supported in several studies by the observation that in the composites, the crack profiles are generally more tortuous (e.g. [16]) and tend to pass around particles rather than cleave them. Taya *et al.* [17] argue that the observed deflection is not always sufficient to explain the entire toughness increase and invoke the added shielding effect of the mean compressive stress in the matrix. A study has recently been reported of a SiC–16 vol% TiB_2 composite [18] which is representative of this type of composite. The particle size of the boride particles was $4.5 \mu m$ which is estimated to be below that which would give spontaneous microcracking during cooling from the processing temperature. After application of a 250 MPa bending stress a marked fall in residual stress as measured by X-ray diffraction was observed providing indirect evidence that microcracking could contribute significant toughening. Microcracks were observed at the particle/matrix interface. It is interesting to note that the systems exhibiting the greatest thermal mismatch such as the silicon nitride matrix composites are those that show least toughening.

Zirconia toughened ceramics exhibit significant increases in toughness [9, 19–23]. In general, a maximum toughening is observed at between 10 and 15 vol% zirconia where values of K_{Ic} around 10 MN m$^{-3/2}$ can be achieved. For particle sizes greater than about $2 \mu m$ the zirconia is mainly monoclinic and the toughening is achieved by microcracking and may therefore be accompanied by a fall in fracture strength. With particle sizes around $1 \mu m$ the tetragonal structure is partly retained after sintering and leads to significant transformation toughening. It appears that for an optimum effect both transformation and microcracking should occur simultaneously [21]. For both modes of toughening the toughness is often observed to fall again for particle fractions beyond 15%. For microcrack toughening this can be explained by linking of microcracks while transformation toughening is believed to be hindered by the onset of particle agglomeration (inadequate mixing).

The fracture strength of zirconia toughened ceramics is determined by a combination of the toughness, defect structure and the matrix grain size. By use of hot isostatic pressing, strengths in excess of 1000 MPa can be achieved in ZTA since high density is achieved while the alumina grain size is kept around $1 \mu m$ [19, 23]. High strengths can also be achieved in transformation toughening composites by inducing transformation (and therefore compressive stresses) in the surface, for example by grinding. This however reduces the toughness in the surface region.

Finally mention should be made of composites that combine particulate toughening with whisker toughening. Several experimental studies have confirmed the predicted synergistic effect of the combined toughening mechanisms. For example, a toughness of $13.5\ \text{MN}\,\text{m}^{-3/2}$ has been observed in an alumina reinforced with 15% zirconia in combination with 20% SiC whiskers [24] and $10.5\ \text{MN}\,\text{m}^{-3/2}$ in mullite reinforced with 20% zirconia and 20% SiC whiskers [25].

6.3.2   *High-temperature deformation and fracture*

Investigations of the effects of temperature on the toughness and fracture strength have been reported for a number of particle composites and reveal the influence of a several phenomena. Figures 6.9 and 6.10 show toughness and strength respectively as a function of temperature for Al_2O_3–TiC and Si_3N_4–TiC composites. The apparent marked increase in toughness of the silicon nitride composite above 900 °C is frequently observed in ceramics that contain glassy grain-boundary phases. The increase is probably the result of crack-tip blunting caused by a softening of the structure. The toughness increase is characteristically accompanied by a sharp fall in strength due to a weakening of the boundaries. The presence of particles is unlikely to

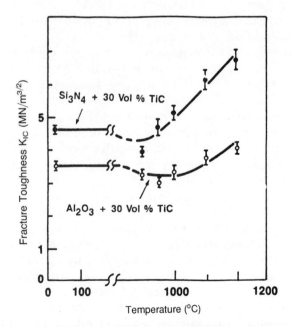

Figure 6.9   Effect of temperature on the fracture toughness of composites of alumina and silicon nitride with 30 vol% TiC additions [26].

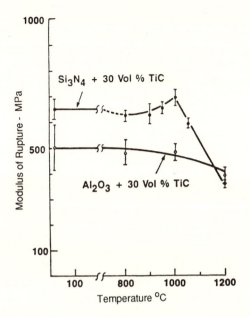

Figure 6.10 Effect of temperature on the fracture strength of alumina and silicon nitride with 30% TiC additions [26].

influence these processes. The alumina matrix composite is affected much less by temperature which is consistent with its much lower content of glassy phase.

In a study of Al_2O_3–25 vol% Ti(N,C) and Al_2O_3–3 vol% ZrO_2 (see Table 6.1), Brandt *et al.* [27] report that from room temperature up to 1000 °C the fracture changes gradually from mixed intergranular/trans-granular to entirely intergranular with an accompanying slight decrease in toughness. At 1200 °C the toughness of the Al_2O_3–ZrO_2 composite fell sharply and considerable sub-critical intergranular crack growth was observed indicating a marked weakening of the boundaries though not associated with a glassy phase. In alumina composites containing higher proportions of transformation toughening zirconia, the toughness is observed to fall continuously with temperature due to a loss of the transformation [28].

The toughness of SiC–16 vol% TiB_2 (Table 6.1) was found to fall slightly with temperature up to 1400 °C [29] while the strength did not change [30]. The insensitivity to temperature probably reflects the fact that only small quantities of additives are present.

Few studies have been made of the creep of particulate composites. Crampon *et al.* [31] report that for silicon nitride (with yttria and alumina sintering aids) containing up to 40 vol% TiN the creep strength decreased and creep rate increased with increasing volume fraction and decreasing particle size of the TiN. The dominant mechanism of creep changed from grain-boundary sliding to cavitation at the particle/matrix boundary. Silicon carbide additions to hot pressed silicon nitride have a similar weakening effect [32]. Some values of hot hardness which give an indication of high-temperature deformation are included in Table 6.1.

6.3.3 *Electrical and thermal conductivity*

Table 6.1 includes values of thermal and electrical conductivity for a number of particulate composites. In general, the thermal conductivity of the composites lie between the values of the individual constituents, consistent with the mixture laws presented in chapter 4. Close agreement with these laws is often not observed probably because of the sensitivity of conductivity to factors other than volume fraction such as internal defects, interface purity, etc.

A dramatic and technically significant effect is the sharp transition from electrical non-conductivity to conductivity when adding particles of a conductor phase to an insulator, for example adding SiC, B_4C, TiN or TiC to alumina or silicon nitride. The transition occurs when the added particles develop a continuous skeleton and the results for particulate composites are consistent with the predicted percolation fraction of 30–40 vol%.

6.3.4 *Oxidation*

The oxidation behaviour of a number of particulate ceramic composites is compared with that of monolithic silicon nitride and silicon carbide in Table 6.2. The products of the oxidation can be gaseous as well as solid.

Table 6.2 Oxidation of ceramics and ceramic composites in air or oxygen

Material	Temperature (°C)	Solid oxidation products	Oxidation rate $(g/m^2 \ h^{1/2})^a$	Ref.
Si_3N_4	1200	SiO_2	0.2–1	–[b]
	1370	SiO_2	1–5	–[b]
SiC	1200	SiO_2	Negligible	[30]
	1370	SiO_2	0.5	[34]
SiC–16 vol% TiB_2	1200	SiO_2, TiO_2	0.07	[30]
SiC–10 vol% Al_2O_3	1370	SiO_2, mullite	3	[34]
Al_2O_3–30 vol% SiC	1375	Mullite, SiO_2	10	[33]
Al_2O_3–45 vol% TiN	1050	TiO_2	63	[37]

[a]Oxidation rate measured in terms of weight loss of composite constituents.
[b]The oxidation of silicon nitride is sensitive to the nature of sintering additives and microstructure; data from several sources.

Therefore to make a meaningful comparison of oxidation rates these are expressed as the weight of composite consumed by the process rather than an overall weight change. The results are expressed in terms of a parabolic rate constant even though for some composites the oxidation deviates from parabolic behaviour after a time, usually as a result of cracking of the oxidation product layer. Since the oxidation of a composite is a multiphase process it is usually complex and differs from system to system. Selected systems will therefore be described individually below.

6.3.4.1 Al_2O_3–SiC composites [33, 34]. Oxidation becomes significant around 1200 °C. The SiC oxidises to SiO_2. This reacts with the alumina to form a metastable aluminosilicate glassy phase and subsequently mullite $(3Al_2O_3.2SiO_2)$. The SiC oxidises more rapidly in the composite than when it is uncombined and this appears to be associated with the penetration of the grain boundaries and interfaces by the fluid glassy phase. The final oxidation product of a composite containing 20 vol% SiC is mullite alone since this amount of silicon corresponds to the stoichiometric composition. At compositions below and above this, excess alumina or silica, respectively, are also present.

6.3.4.2 SiC–16 vol% TiB_2 composite [30, 35]. Measurable oxidation begins at around 1200 °C with the formation of TiO_2 and SiO_2. Boron is thought to disappear as gaseous oxides. The oxidation rate decreases rapidly with time. It appears that the silica quickly forms a protective film and the oxidation does not penetrate to boride particles below the surface.

6.3.4.3 Al_2O_3–TiC [36] and Al_2O_3–TiN [37] composites. Significant oxidation begins as low as around 800 °C with the formation of rutile (TiO_2). In the TiC composite at higher temperatures, the titanium oxide was observed to react with the alumina to form aluminium titanate (Al_2TiO_5) and a similar reaction probably occurs for TiN composites. Both rutile and aluminium titanate exhibit a volume increase relative to the substrate and therefore begin to crack up with a consequent increased oxidation rate.

6.4 Wear properties and applications

The most important applications of particulate composites are those involving some form of wear resistance, for example wear parts, knife blades and cutting tools. As is indicated by the discussion of wear mechanisms in chapter 4, improvements in wear resistance achieved by adding one or more ceramic phases to a ceramic matrix depend very much on the details of the wear situation, for example the geometry and composition of the wearing surfaces involved, the environment, the mechanical loads, etcetera. This can be illustrated by specific examples reported in the literature.

Sarin *et al.* [38] have shown for a series of particulate composites that resistance to abrasive wear (abrasion by a 45 μm diamond dry abrasive in argon) could to a first approximation be related to the the hardness and toughness through the parameter $K_{Ic}^{3/4} \times H^{1/2}$ (see Figure 6.11). Although this parameter has some basis in a model involving microfracture, it should in general be regarded merely as a figure of merit giving only a rough relative indication of wear resistance in abrasive situations. Buljan and Sarin [39] suggest that in some cases this figure of merit may also be applied to metal cutting; however when high temperatures are generated, chemical and dissolution wear dominate. Values for the figure of merit of several ceramic composites are included in Table 6.1 and confirm the improvement brought about by additions of particles to ceramics in several cases.

Holz *et al.* [40] examined the abrasive wear resistance (180 and 600 grit SiC abrasive) of a large range of ceramics and ceramic composites; selected examples are given in Table 6.3. In this study no clear relationship was seen between wear resistance and a hardness/toughness combination. The wear resistance of a ceramic was usually improved more by increasing the hardness, than the toughness. Thus the wear

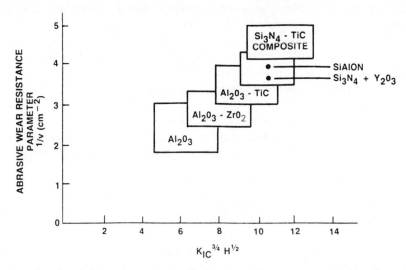

Figure 6.11 Relationship between abrasion wear resistance of particulate composite cutting tools and their combined toughness and hardness [38]. K_{Ic} = fracture toughness; H = hardness.

Table 6.3 Abrasive wear resistance of ceramic and ceramic composites (pin on 15 μm SiC paper, 2.2 MPa, 300 mm/min)) [40]

Material	Wear rate (volume per unit area and unit length)
Al_2O_3	5–10
Hot-pressed β-sialon	4
Si_3N_4-Ti(C,N)	10
Mullite	28
Mullite–20% ZrO_2	20
Mullite–10% ZrO_2–20% SiC_w	6.5
Al_2O_3–ZrO_2	6
Al_2O_3–ZrO_2–Ti(C,N)	2.3
Al_2O_3–SiC (platelets)	8.2

resistance of alumina was improved more effectively by additions of TiC, TiN and SiC than by additions of zirconia. It was also noted that if a second phase is too poorly bonded to the matrix it can lead to increased wear rates by being pulled out and forming abrasive debris. A similar effect can presumably arise if internal mismatch or transformation stresses lead to excessive microcracking.

A number of ceramic materials exhibit reductions in friction at elevated temperatures due to the formation of lubricating surface layers. This effect could be exploited to introduce *in situ* lubricants, designed for specific temperature ranges, into ceramic composites. For example, the friction of silicon carbide is found to decrease dramatically above 1000 °C in some situations due to the development of a thin graphitic surface layer [41]. Oxides of titanium have low coefficients of friction indicating that particles of titanium compounds in composites could provide lubricant potential.

The advantage of ceramics compared to other tool materials in metal cutting applications is that the temperature at the tool/workpiece interface can be raised permitting increased removal rates [42]. The applicability of the first ceramic tools based on monolithic alumina was limited by brittle fracture caused by local thermal stresses, thermal shock and impact loading inherent to machining. The successful mixed ceramic composites exhibit a reduced tendency to fracture presumably because of increased thermal conductivity and, in some cases, improved toughness/strength. A guide to this improvement is provided by thermal shock resistance figures of merit, an example being included in Table 6.1.

An important area of application for ceramic tools is the machining of cast iron, for example in brake drums and discs (see Figure 6.12). Alumina based composites are preferred for high-quality cast iron with low slag content and with the rough as-cast surface removed, while silicon nitride based composites are better for poorer qualities since they are more resistant to the wear caused by slag and sand inclusions [42].

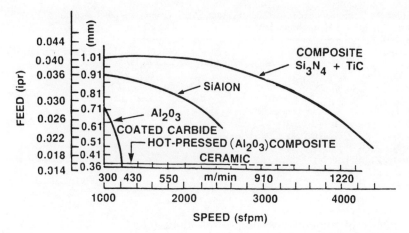

Figure 6.12 Comparison of the application range limits of various ceramics in machining of gray cast iron [38].

Ceramic tools can be applied to tool steels and hardened steels but are less successful for machining of other steels; the thermo-mechanical loads lead to fracture of oxide materials while the life of composites containing silicon compounds is severely limited by chemical wear effects. Some improvements have been achieved by adding alumina to a Si_3N_4–TiC composite [39] thus reducing chemical effects.

Alumina based particulate ceramics exhibit good wear resistance in machining of high-temperature Ni-base alloys but again their use is restricted by fracture [43, 44]. This can be avoided by the combination of particle and whisker reinforcement. Exner et al. [45] studied the performance of a range of alumina based composites in cutting Inconel 718. The composites contained combinations of up to 30% zirconia and 45% SiC whiskers giving toughness in the range 6.5–8.2 MN m$^{-3/2}$ and hardness in the range 19–21.5 GPa. It was found that resistance to flank wear increased with hardness while resistance to depth-of-cut notch wear increased with fracture toughness.

6.5 Conclusions

The benefits of creating composites by incorporating particles of one ceramic phase in the matrix of another are well proven and a number of particulate composites have been developed commercially. An important advantage of particle reinforcement is that the composites can be produced by conventional ceramic processing techniques. The main effects of reinforcement are improvements in toughness (in particular by

use of microcrack and transformation toughening with zirconia particles), hardness (using carbides, nitrides and borides) and thermal
conductivity (using carbides, nitrides and borides). These improvements
have opened up a range of applications involving wear resistance, in
particular metal cutting. The possibility of controlling the electrical
properties of ceramics by a suitable combination of phases has led to the
development of certain particulate composites as electrical conductors
for severe, high temperature environments.

The possible combinations of ceramics in particulate ceramics is
almost limitless and consequently the potential for their future development is very large.

References

1. Marple, B. R. and Green, D. J. (1989) Presented at *91st Annual Meeting of the American Ceramic Society*, Indianapolis, paper 1–SI–89.
2. Crudele, S. D. and Kriven, W. M. (1989) Presented at *91st Annual Meeting of the American Ceramic Society*, Indianapolis, paper 2–SI–89.
3. Biel, J. P., Aghajanian, M. K. and Anderson, C. A. (1989) Presented at *91st Annual Meeting of the American Ceramic Society*, Indianapolis, paper 16–SI–89.
4. Rice, R. W. and Freiman, S. W. (1981) *J. Am. Ceram. Soc.* 64, 350–354.
5. Sarin, V. K. (1988) *Proc. 3rd Int. Conf. on Science of Hard Materials*, ed. Sarin, V. K., Elsevier, London, pp. 151–159.
6. Kang, E. S., Jang, C. W., Lee, C. H., Kim, C. H. and Kim, D. K. (1989) *J. Am. Ceram. Soc.* 72, 1868–1872.
7. Richerson, D. W. (1982) *Modern Ceramic Engineering*, Marcel Dekker, New York.
8. Bostedt, E., Persson, M. and Carlsson, R. (1989) *Proc. 1st European Ceramic Soc. Conf.—Euro-Ceramics vol. 1*, eds. deWith, G., Terpstra, R. A. and Metselaar, R., Elsevier, London, pp. 1. 140–1. 144.
9. Butler, E. P. (1985) *Mater. Sci. Technol.* 1, 417–432.
10. Rice, R. W. (1984) *MRS Symposium Proc. vol. 32*, eds. Brider, C. J., Clark, D. E. and Ulrich, D. R., North-Holland, Amsterdam, pp. 337–345.
11. Newkirk, M. S., Lesher, H. O., White, D. R., Kennedy, C. R., Urquart, A. W. and Claar, T. D. (1987) *Ceram. Eng. Sci. Proc.* 8, 879.
12. Janney, M. A. (1986) *Am. Ceram. Soc. Bull.* 65, 357–362.
13. Buljan, S. T. and Sarin, V. K. (1984) *Sintered Metal Ceramics Composites*, ed. Upadhyaya, G. S., Elsevier, Amsterdam, pp. 455–468.
14. Buljan, S. T. and Sarin, V. K. (1987) *Composites* 18, 99–106.
15. Buljan, S. T., Baldoni, J. G. and Huckabee, M. L. (1987) *Am. Ceram. Soc. Bull.* 66, 347–352.
16. Baldoni, J. G., Huckabee, M. L. and Buljan, S. T. (1986) *Tailoring Multiphase and Composite Ceramics*, eds. Tressler, R., Messing, G., Pantano, C. and Newnham, R., Plenum Press, New York, pp. 329–344.
17. Taya, M., Hayashi, S., Kobayashi, A. S. and Yoon, H. S. (1990) *J. Am. Ceram. Soc.* 73, 1382–1391.
18. Magley, D. J., Winholtz, R. A. and Faber, K. T. (1990) *J. Am. Ceram. Soc.* 73, 1641–1644.
19. Claussen, N. (1984) *Advances in Ceramics* 12, 325–351.
20. Becher, P. F. (1986) *Acta Metall.* 34, 1885–1891.
21. Kosmac, T., Swain, M. V. and Claussen, N. (1985) *Mater. Sci. Eng.* 71, 57–64.
22. Rühle, M., Evans, A. G., McMeeking, R. M., Charalambides, P. G. and Hutchinson, J. W. (1987) *Acta Metall.* 35, 2701–2710.
23. Shin, D. W., Orr, K. K. and Schubert, H. (1990) *J. Am. Ceram. Soc.* 73, 1181–1188.

24. Becher, P. F. and Tiegs, T. N. (1987) *J. Am. Ceram. Soc.* 70, 651–654.
25. Claussen, N. and Petzow, G. (1986) *Am. Ceram. Soc. Bull.* 66, 649–662.
26. Baldoni, J. G., Buljan, S. T. and Sarin, V. K. (1986) in: *Proc. 2nd Int. Conf. Sci. Hard Materials*, eds. Almond, E. A., Brookes, C. A. and Warren, R., Inst. Physics/Adam Hilger, Bristol, pp. 427–458.
27. Brandt, G., Johannesson, B. and Warren, R. (1988) *Mater. Sci. Eng.* A105/106, 193–200.
28. Orange, G., Fantozzi, G., Homerin, P., Thevenot, F., Leriche, A. and Cambier, F. (1986) *Proc. 2nd Int. Symp. Ceramic Materials and Components for Engines*, eds. Bunk, W. and Hausner, H., DKG, Bad Honnef, BRD, pp. 567–574.
29. Jenkins, M. G., Salem, J. A. and Seshadri, S. G. (1989) *J. Composite Mater.* 23, 77–91.
30. McMurty, C. H., Boecker, W. D. G., Seshadri, S. G., Zanghi, J. S. and Garnier, J. E. (1987) *Am. Ceram. Soc. Bull.* 66, 325–329.
31. Crampon, J., Duclos, R. and Vivier, P. (1989) *Proc. 1st European Ceramic Soc. Conf.*, *Vol. 3*, eds. deWith, G., Terpstra, R. A. and Metselaar, R., Elsevier, London, pp. 3.298–3.302.
32. Birch, J. M. and Wilshire, B., (1978) *J. Mater. Sci.* 13, 2627–2636.
33. Luthra, K. L. and Park, H. D. (1990) *J. Am. Ceram. Soc.* 73, 1014–1023.
34. Kossowsky, R. and Singhal, S. C. (1975) *Grain Boundaries in Engineering Materials*, eds. Walter, J. L., Westbrook, J. H. and Woodford, D. A., Claitors Publishing Division, Baton Rouge, pp. 275–287.
35. Seshadri, S. G., Srinivasan, M., MacBeth, J. W. and Ten Eyck, M. O. (1989) *Proc. 3rd Int. Symp. Ceramic Materials and Components for Engines*, pp. 1419–1428.
36. Borum, M. P., Brun, M. K. and Szaler, L. E. (1988) *Adv. Ceram. Mater.* 3, 491–497.
37. Mukerji, J. and Biswas, S. K. (1990) *J. Am. Ceram. Soc.* 73, 142–145.
38. Sarin, V. K., Buljan, S. T. and Smith, J. T. (1985) *Science and Technology*, ed. Parker, S. P. McGraw-Hill, New York, pp. 441–449.
39. Buljan, S. T. and Sarin, V. K. (1986) in: *Proc. 2nd. Int. Conf. Sci. Hard Materials*, eds. Almond, E. A., Brookes, C. A. and Warren, R., Inst. Physics/Adam Hilger, Bristol, pp. 873–882.
40. Holz, D., Janssen, R., Friedrich, K. and Claussen, N. (1989) *J. European Ceram. Soc.* 5, 229–232.
41. Buckley, D. H. and Rabinowicz, E. (1986) *J. Am. Ceram. Soc.* 73, 825–849.
42. Brandt, G. (1990) *Encyclopedia of Materials Science and Engineering, supplementary vol. 2*, ed. Cahn, R. W., Pergamon, Oxford, pp. 761–765.
43. Brandt, G. (1990) *J. European Ceram. Soc.* 6, 273–290.
44. Billman, E. R., Mehrotra, P. K., Shuster, A. F. and Beeghly, C. W. (1988) *Ceram. Eng. Soc. Proc.* 9, 543–552.
45. Exner, E. L., Jun, C. K. and Moravansky, L. L. (1988) *Ceram. Eng. Soc. Proc.* 9, 597–602.

7 Long-fibre reinforced ceramics

D. C. PHILLIPS

7.1 Introduction

This chapter is concerned with the manufacture and properties of long-fibre composites. These are materials containing relatively large volume fractions of continuous aligned fibres either as a unidirectional composite or in laminae, the laminae being stacked together and consolidated to produce a multidirectional laminate. The technology is most advanced for reinforced glass and glass-ceramic matrix systems in which a slurry impregnation route is employed to incorporate the fibres into the unconsolidated matrix, followed by hot pressing for consolidation, and most of this chapter is devoted to these materials [1–4]. Most development work has been carried out using multifilamentary tows of small diameter (7–15 μm) SiC or C fibres. In principle a similar approach may be used for other ceramic matrix systems but for a variety of reasons this has not yet been accomplished to a significant an extent [2, 3, 5]. Other routes which are also considered briefly in this chapter include reaction bonding, to produce Si_3N_4 composites, and gaseous reaction of metal matrices, to produce for example mixed Al_2O_3/Al matrix composites. Neither of these is at as advanced a stage of development as the glass and glass-ceramic composites.

Hot-pressed glass and glass-ceramic matrix composites are the highest strength ceramic-matrix composites available and they also exhibit high values of toughness. For example, flexural strengths of around 1600 MPa have been obtained for unidirectional SiC fibre reinforced borosilicate glass composites containing 60 vol% of fibre with works of fracture of around 70 kJ m^{-2} and candidate K_Q values of ~ 25 MPa m$^{1/2}$ [6]. These high strengths are obtained through producing matrices with negligible porosity while at the same time minimising damage to the fibres during fabrication of the composite. In their mechanical performance the properties of these composites are similar in many respects to high performance advanced polymer composites.

These materials have considerable potential value for high-temperature applications where high strength, high toughness, low strength variability and low weight are required. These include components of advanced aircraft engines and aerospace applications where atmospheric friction heating is a problem. They may also have applications in armour. They can be designed to have zero thermal expansion coefficients, which can make them valuable for a range of applications such as in optical equipment. Compared with polymer composites they absorb negligible quantities of water and gases and thus present less problem of degassing in space applications.

There are several reasons why glasses and glass ceramics have provided the most useful matrices for composites of this type [2, 3]. Firstly, the glasses provide a choice of a wide range of coefficients of thermal expansion (CTE); they permit relatively low processing (hot pressing) temperatures compared with true ceramics, and they have lower elastic moduli than engineering ceramics. These points are discussed below. Table 7.1, adapted from [4], shows some glasses and glass ceramics of current interest for composites.

As indicated in chapter 4, the CTE of the matrix is important since if it differs from that of the fibre, mismatch strains will be set up in the two phases during cooling from the fabrication temperature which is typically around 1000 °C or higher. If the CTE of the matrix (α_m) is higher than that of the fibre (α_f), the tensile stress set up in the matrix may be sufficient to cause extensive matrix cracking. It is for this reason that only a relatively small group of the large number of possible fibre/matrix combinations that are available can produce useful composites [3–5]. Ideally the CTE of the fibre should be equal to or slightly higher than that of the matrix. Having α_f higher than α_m can be beneficial in two respects. Firstly, in a unidirectional fibre composite, the matrix will be placed in residual compression parallel to the fibres which will result in enhancement of its load-induced cracking strain. For composites in which the matrix failure strain is lower than that of the fibre and which therefore begin failing by multiple matrix microcracking (the majority of glass-matrix composites are of this type), the increased matrix failure strain will give improvements in performance. Multiple matrix microcracking occurring at relatively modest operating stresses is described in chapter 4 and its implications are discussed in more detail in section 7.4.4. The second effect when $\alpha_f \gtrsim \alpha_m$ is that some radial shrinkage of the fibre from the matrix can occur, which can result in a decrease in the fibre–matrix bond strength and a resulting improvement in toughness and strength, as demonstrated in Table 7.2 for carbon fibre systems [7] and Figure 7.1 for a SiC fibre reinforced cordierite system [8]. Table 7.2 shows work of fracture (WoF) and interlaminar shear strength (ilss) data for composites consisting of carbon fibres in borosilicate glass and lithium–alumino–silicate (LAS) glass-ceramic matrices. The carbon fibres had a radial CTE of 8×10^{-6} °C^{-1} and the borosilicate and LAS had CTEs and stress-relieving temperatures, respectively, of around 3.5×10^{-6} °C^{-1} and 500 °C and 2×10^{-6} °C^{-1} and 1000 °C. The relative radial shrinkage was much greater for the LAS

Table 7.1 Some glass and glass-ceramic matrices of current interest (after Prewo and Brennan [4])

Matrix type	Major constituents	Minor constituents	Major crystalline phase	Maximum use temp. (°C) (in composite form)
Glasses				
Borosilicate	B_2O_3, SiO_2	Na_2O, Al_2O_3	–	600
Aluminosilicate	Al_2O_3, MgO, CaO, SiO_2	B_2O_3, BaO	–	700
High silica	SiO_2	B_2O_3	–	1150
Glass ceramics				
LAS-I	Li_2O, Al_2O_3 MgO, SiO_2	ZnO, ZrO_2 BaO	β-Spodumene	1000
LAS-II	Li_2O, Al_2O_3 MgO, SiO_2, Nb_2O_5	ZnO, ZrO_2 BaO	β-Spodumene	1100
LAS-III	Li_2O, Al_2O_3 MgO, SiO_2, Nb_2O_5	ZrO_2	β-Spodumene	1200
MAS	MgO, Al_2O_3, SiO_2	BaO	Cordierite	1200
BMAS	BaO, MgO, Al_2O_3, SiO_2	–	Barium osumillite	1250
Ternary mullite	BaO, Al_2O_3, SiO_2	–	Mullite	~1500
Hexacelsian	BaO, Al_2O_3, SiO_2	–	Hexacelsian	~1700

Table 7.2 The effect of radial thermal expansion mismatch on the interlaminar shear strength (ilss) and work of fracture (WoF) of two different types of carbon fibre borosilicate glass and lithium aluminosilicate glass-ceramic composites (after Phillips [2])

	Relaxation temperature	CTE $(10^{-4}\ °C^{-1})^a$	Radial shrink $(°C\ 10^{-8}\ m)$	Flexural strength (MPa)	ilss (MPa)	WoF $(kJ\,m^{-2})$
Borosilicate glass and high modulus fibre	520	3.5	0.9	459	59	3.1
Borosilicate glass and high strength fibre	520	3.5	0.9	575	71	3.6
Glass-ceramic and high modulus fibre	1000	2.0	2.4	558	32	4.5
Glass-ceramic and high strength fibre	1000	2.0	2.4	574	26	10.3

^{a}C fibre CTE: radial $8 \times 10^{-6}\ °C^{-1}$; axial ≈ 0.

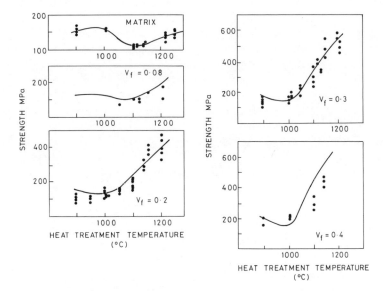

Figure 7.1 The strength of composites consisting of SiC fibre in a cordierite glass-ceramic matrix, hot-pressed at 900 °C and then modified by heating for 1 h at the indicated temperature (after Aveston [8]).

system than the borosilicate system resulting in a lower bond strength and thus a lower ilss and higher WoF. Figure 7.1 shows how the room temperature strength of SiC reinforced cordierite, hot-pressed at 900 °C, varied with heat treatment temperature. On heat treatment below 1000 °C the matrix consisted of μ-cordierite which has a higher thermal expansion coefficient than SiC fibre, so that in the composite the matrix is in tension and the fibre under radial and axial compression. These composites were relatively weak and brittle, with planar fracture surfaces due to the strong bond between fibre and matrix. After treating above 1000 °C the μ-cordierite is changed into high cordierite which has a lower expansion coefficient than the fibres. The matrix is thus in compression parallel to the fibres and radially the fibre contracts from the matrix producing a weaker interface. These composites were much stronger and tougher displaying fracture surfaces with protruding fibres.

Figure 7.2 shows the variation of viscosity of some glasses with temperature. Much work has been carried out on fibre reinforced borosilicate glasses because of the advantageously low CTE of these systems. The softening point of such systems is around 800 °C and good quality composites with low porosity matrices might be expected to be produced by hot pressing at temperatures below 1000 °C. However problems can arise due to the nucleation and growth of crystalline phases such as cristobalite and this has to be taken into account in optimising the fabrication conditions [9]. Glass-ceramic matrices might be produced either by consolidating a glass matrix and then heat-treating to devitrify into a glass ceramic (ceramming), or by hot pressing a

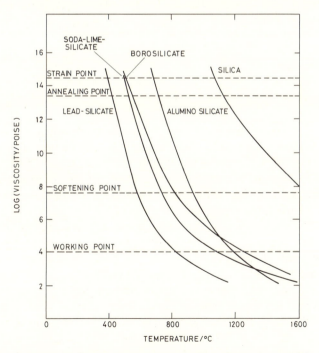

Figure 7.2 The viscosity of some glasses as a function of temperature (after Holloway [54]).

glass-ceramic powder. Sol-gel routes have also been explored to reduce heat treatment temperatures. In general, processing temperatures by these different routes tend to lie in the range 900–1200 °C for glass-based systems, as opposed to > 1600 °C for hot-pressed engineering ceramics [5].

The relatively low elastic moduli of glass based matrices is valuable because the reinforcement efficiency of a fibre in a matrix depends on the ratio of the moduli of fibres and matrix, E_f/E_m. Typical Young's moduli of fibres lie in the range from 200 GPa for SiC fibres produced from polymeric precursors to 360 GPa for high modulus PAN based carbon fibres. Young's moduli of glasses and glass ceramics lie typically in the range 60–90 GPa, while those of engineering ceramics are typically 250–400 GPa. Thus better load transfer occurs in a glass or glass-ceramic matrix which for a given matrix stress level implies greater strengthening.

The main disadvantage of these matrices is that they exhibit a lower temperature capability than engineering ceramics, which restricts their use through creep, for glasses to ≈ 600 °C and glass ceramics to $\approx 1000–1200$ °C. However, as will be discussed later, this is not currently the main restriction on the high-temperature use of these composites as other problems, primarily to do with the stability of existing fibres, produce more severe problems.

7.2 Fabrication

7.2.1 *Glass and glass-ceramic composites*

The manufacture of glass and glass-ceramic composites by the slurry impregnation route involves three stages: the production of a pre-preg tape of fibre and unconsolidated matrix material; cutting and lay-up of the pre-preg tape to an appropriate pattern for the required shape of component and stacking sequence of plies; and hot pressing to produce the final, consolidated composite component.

Figure 7.3 shows diagrammatically some key features of the slurry impregnation route [1, 2]. The continuous, multifilament tow, which contains typically between 1000 and 10 000 fibres depending on its type, may contain a twist which can be conveniently released by passing the tow up a column. The tow may then be spread, for improved impregnation. In Figure 7.3 this is achieved by counter currents of air through fan tail jets. The tow is then passed through a slurry consisting of powdered matrix material in a solution of solvent and organic binder. The powdered matrix material can be produced by conventional comminution processes employed in ceramics technology and is typically

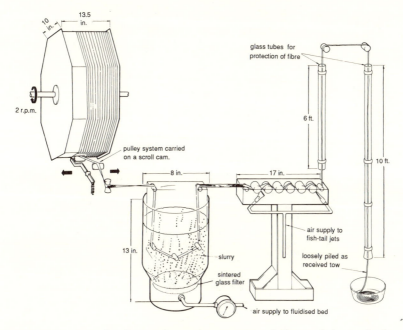

Figure 7.3 Apparatus for the continuous impregnation of fibre and the production of pre-impregnated tape [1].

$\sim 20\text{-}40~\mu\text{m}$ in size, but with a proportion of finer powder [2]. The slurry needs to be agitated during impregnation in order to keep the powder dispersed. In Figure 7.3 this is achieved by air agitation through a sintered glass disc. The tow leaves the slurry and is wound onto a drum, the solvent evaporating away to leave a sheet consisting of fibres within which the powder is intimately intermingled, all bound together with binder. The handleable sheet is then removed from the drum and can be cut to appropriate sizes for the next stage of the process.

A variation on this route is to use a gel, produced by sol-gel chemistry, instead of a slurry. The problem with this is that very large shrinkages occur during consolidation leading to poor composites due to shrinkage cracks and voidage. An alternative, more successful, technique is to disperse powdered matrix material in a gel of a similar material, the gel then acting both as a binder and as a source of matrix material [10]. This has been claimed to lead to improvements in processing partly by avoiding the need to burn off the binder. The slurry impregnation route is capable of good reproducibility and control as shown in Figure 7.4 in which the volume fraction of fibre in a composite is shown as a function of the composition of the slurry [2].

After being cut to appropriate sizes the laminae are laid up in a die for hot pressing, in a method similar to conventional polymer composite

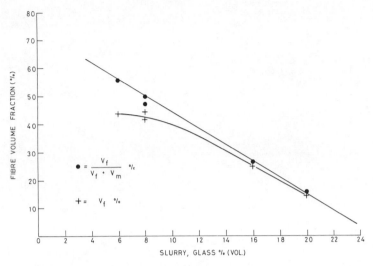

Figure 7.4 The dependence of the volume fraction of fibres in the composite on the composition of the slurry for a carbon fibre reinforced glass ceramic [2].

pre-preg technology with the difference that the hot-pressing temperatures are very much higher. Graphite dies are particularly suitable for this purpose and bolstered graphite dies can be used for large components. Figure 7.5 illustrates one specific die assembly which has been used for this purpose [2].

The hot-pressing cycle has to be optimised by taking into account a number of requirements. The pre-preg has to be heated at a sufficiently slow rate to burn off all the binder. At an early stage a low pressure may need to be applied to partially consolidate the preform in order to prevent segregation of powder as the binder is removed, however the pressure must be sufficiently low to avoid fibre damage. The temperature is then raised to an adequately high temperature to soften the powder sufficiently to allow a high pressure to be applied to consolidate the matrix, again without damaging the fibres. The compact is then cooled under pressure to a low enough temperature for the pressure to be released without spring-back of the fibres or release of dissolved gas from the matrix, in order to maintain a low porosity. Typical maximum temperatures and pressures are around 1200 °C and 6 MPa. A fuller account for a carbon fibre reinforced borosilicate glass system is given in [2]. Figure 7.6 shows some prototype components made in this way illustrating the complexity of shape possible by hot pressing [11].

In polymer composites technology a wide range of fabrication techniques have been developed to suit particular production requirements such as complexity of shape or the need for continuous processing. The

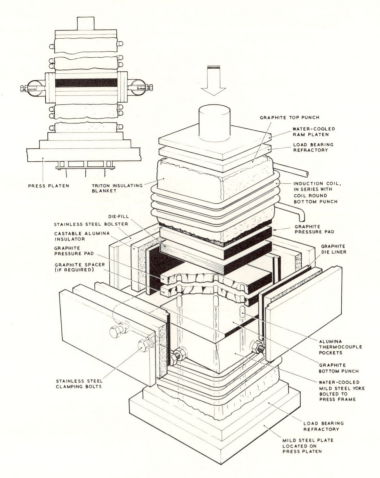

GRAPHITE TOP PUNCH

WATER-COOLED
RAM PLATEN

LOAD BEARING
REFRACTORY

INDUCTION COIL,
IN SERIES WITH
COIL ROUND
BOTTOM PUNCH

GRAPHITE
PRESSURE PAD

GRAPHITE
DIE LINER

ALUMINA
THERMOCOUPLE
POCKETS

GRAPHITE
BOTTOM PUNCH

WATER-COOLED
MILD STEEL YOKE
BOLTED TO
PRESS FRAME

LOAD BEARING
REFRACTORY

MILD STEEL PLATE
LOCATED ON
PRESS PLATEN

PRESS PLATEN

TRITON INSULATING
BLANKET

DIE-FILL

STAINLESS STEEL BOLSTER

CASTABLE ALUMINA
INSULATOR

GRAPHITE
PRESSURE PAD

GRAPHITE SPACER
(IF REQUIRED)

STAINLESS STEEL
CLAMPING BOLTS

Figure 7.5 A large die assembly used for hot pressing fibre reinforced glass-ceramic plates
[1].

attractiveness of these techniques has not been lost on ceramic technolo-
gists who have attempted to adapt the processes to glass and glass-
ceramic composites. However Figure 7.2 indicates a major problem in
attempting these developments. Filament winding, for example, is used
routinely for liquid resin matrices such as epoxide resins, to produce
both axisymmetric and non-axisymmetric structures [12], but such resins
have low viscosities typically ≈ 2–6 P. To achieve viscosities as low as
this with glasses would require very high temperatures, $> 1200\,°C$.
Although some modest success has been reported in filament winding
glass-matrix composites at elevated temperatures other, related possibilities are also being explored including an adaptation of the powder slurry

(a)

(b)

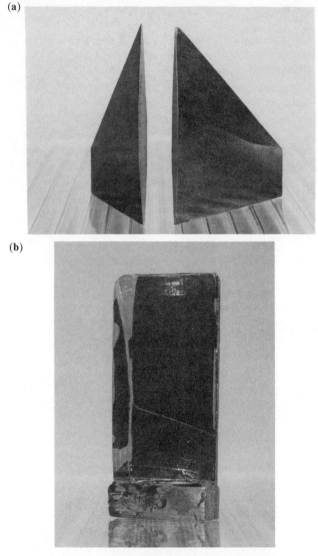

Figure 7.6   (a) Prototype fin and (b) turbine blade manufactured from SiC fibre reinforced borosilicate glass (courtesy of R. W. Davidge).

impregnated tape process described above, and the use of liquid chemical precursors to glasses and glass ceramics [13]. An adaptation of the slurry impregnation route to produce pultruded, continuous, constant-section stock has also been developed [14] and a procedure employing molten glass has been developed for matrix transfer moulding [4].

7.2.2 *Polycrystalline ceramic composites*

A number of different routes are being explored for the development of true, high-temperature, ceramic-matrix composites including the slurry impregnation, hot-pressing route [5]. The major problem is that conventional hot-pressing techniques for engineering ceramics require very high temperatures in excess of 1600 °C. For example hot-pressed Si_3N_4 composites required 1700–1800 °C and SiC composites 2100 °C [15–17]. At these temperatures the currently available fibres tend to degrade either because of intrinsic instability of the fibre at temperature, as in the case of SiC-type fibres produced from polymer precursors, or because of reaction with the matrix. For this reason a major objective of current research is to reduce processing temperatures and/or times. Processes which can achieve this include chemical vapour infiltration described in chapter 8, sol-gel techniques, reaction bonding and gaseous reaction of metal matrices.

Reaction bonding has been investigated for producing Si_3N_4-matrix composites at temperatures lower than those necessary for hot pressing [18–20]. The advantage of reaction bonding over pressureless sintering for CMCs is that consolidation occurs by filling the void space within the green part with reaction product, rather than through shrinking [20]. For example, composites of Nicalon SiC fibres in a Si_3N_4 matrix with a density of 70–80% have been produced by rapid nitridation of Si/Si_3N_4 powder in 3–4 h at 1350 °C [18, 19]. Promising results were obtained but the high porosity of the matrix and some degradation of the fibres resulted in composites with only modest properties. Another promising approach, still at an early stage of development, is the so-called Lanxide DIMOX^{TM} process [21–23]. In this process the fibres are immersed in molten metal which is then exposed to a gaseous reactant. At the appropriate temperature, in the range 900–1350 °C, a rapid reaction takes place with the formation of a reaction product which grows outward from the original metal surface. The resulting matrix is a mixture of unreacted metal and ceramic such as aluminium/aluminium oxide or aluminium/aluminium nitride. It is reported that fibre composites of Al_2O_3 fibres and SiC fibres in Al_2O_3/Al matrices have been produced in this way.

7.3 Interphase reactions

In the light of problems which emerged in later work with oxide fibres and non-stoichiometric SiC fibres, it was fortunate that the early work on the development of glass and glass-ceramic composites, which occurred in the late 1960s and early 1970s, was carried out mainly with carbon fibres [1–3] and with large diameter stoichiometric, SiC fibres [8] produced by a chemical vapour deposition (CVD) route. In the early 1970s interest waned in the carbon fibre composites because the composites oxidised at modest temperatures ($\approx 400\,°C$) in air, and in the CVD SiC fibre composites because of their intrinsically high cost, limited availability and the difficulties associated with processing composite systems containing large diameter, monofilament reinforcement. An upsurge in enthusiasm for ceramic composites began in the late 1970s with the increasing availability of other ceramic fibres, notably the multifilamentary, small diameter, SiC-type fibre produced by the Yajima process but also oxide fibres [2, 4]. Most of the ensuing development work has concentrated on the SiC-type fibres produced from polymer precursors by Nippon Carbon (Nicalon) and Ube Industries (Tyranno) rather than oxide fibres because of *inter alia* problems of reactive bonding between oxide fibres and oxide-glass matrices, which results in a relatively brittle, low-strength composite [24], and of creep of oxide fibres. The SiC fibres offered the promise of glass-ceramic matrix composites which could be stable to high temperatures in air.

However, development programmes on SiC fibre glass-ceramic matrix composites during the 1980s revealed troublesome problems in attempts to manufacture composites with good, consistent properties that are retained with time at temperature; problems which had not been apparent with the earlier carbon fibre composite systems. These resulted mainly from the non-stoichiometry of the fibres which led to instabilities in their compositions at the elevated temperatures of composite manufacture and to reaction between the fibres and the matrix. Other problems included the formation of harmful crystalline phases in the glass-matrix systems, which occurred because of attempts to reduce processing temperatures both in order to limit degradation of fibres and in order to improve the economics of composite manufacture, and problems in obtaining good glass-ceramic structures. Optimisation of the processing route for SiC fibre reinforced glasses and glass ceramics, the acquisition of reproducibility, and transfer of technology to large scale manufacture thus proved more difficult than for carbon fibre systems. Details of manufacturing schedules for composites of this type are not readily available for commercial reasons and so only rather general comments are possible.

Nicalon (Si–C–O) and Tyranno (Si–Ti–C–O) fibres are non-stoichiometric. Nicalon fibre for example contains approximately 59 wt% Si, 31 wt% C and 10 wt% O [25], an excess of C and O of approximately 10 at% each. It has been demonstrated that in order to produce strong, tough Nicalon fibre composites, fabrication conditions are needed such that a relatively weak, carbon rich, fibre–matrix interface, about 100–400 Å thick is formed [4, 26]. This provides good toughness due to crack deflection and fibre pull-out. Under fabrication conditions such that the carbon rich interfacial zone is not developed, the resulting composites are weak and brittle [26]. Detailed electron microscopy studies [27] have shown that the carbon rich interface is a cryptocrystalline carbon/graphite reaction layer between the matrix and fibre. It has been suggested that the chemistry of the fibres can be characterised by compositional activities of SiC, SiO_2 and C, each of which are approximately equal to unity, and that the morphology and chemistry of the interface are consistent with the interface having been formed by an SiC oxidation reaction which is rate-limited by solid-state diffusion of silicon and oxygen down a silica activity gradient from fibre to matrix. The evolution and stability of the graphitic interface depends upon the composition of the matrix and in particular its basicity, which affects both the activity of the silica and the oxidation potential of the matrix [27]. Prolonged exposure to temperature can lead to interdiffusion of species between the fibre and matrix, a process which is probably aggravated by the non-stoichiometry of the fibres.

In order to produce a good quality composite the processor needs therefore to keep the temperature of processing as low as is feasible but this might, in principle, lead to mechanical damage to fibres because of the hardness of the particles. In addition the possibility of growth of unwanted phases must be taken into account. Recent electron and optical microscopy and related studies have begun to reveal features of the development of microstructure in glass-ceramic matrices [4] and in glass matrices [9, 28, 29]. Even a relatively simple matrix system such as borosilicate glass can yield problems. Early work on borosilicate glass matrices employed relatively high temperatures (1200 °C) and low pressures (4 MPa) and yielded good quality composites [2]. Later work attempted to employ lower temperatures (~ 850 °C) partly to reduce any problems of degradation of SiC fibres and partly for easier processing, but these lower temperatures resulted in the formation of the high CTE phase, cristobalite, resulting in substantially mircocracked matrices and lower strength composites, as shown in Table 7.3 for carbon fibre borosilicate glass [9]. It has also been suggested that the presence of SiC

Table 7.3 Effect of fabrication conditions on a carbon-borosilicate glass composite (after Davies *et al.* [9])

Fabrication route	Flexural strength (40 mm span) (MPa)	Matrix condition
850 °C, 12 MPa, 5 min	533 ± 110	Microcracked, much cristobalite present
1200 °C, 4 MPa, 1 min	1226 ± 52	Glassy, negligible cristobalite present

fibres can enhance the crystallisation of glass [28]. The solution to the conflicting problems faced by the manufacturer is not trivial and requires a careful understanding of the relationship between the different phases possible in a fibre–matrix system.

7.4 Mechanical properties

7.4.1 *Introduction*

There have been two, main, rather different reasons for the development of ceramic-matrix composites. One has been to develop a system which has the properties, essentially, of the monolithic ceramic matrix but with a significantly enhanced toughness; composites in this category include the whisker toughened and particulate toughened ceramics. The other has been to attempt to utilise, to the full, the high strengths and stiffness offered by continuous ceramic fibres, at as high a temperature as possible. This second class of composites, which includes the materials described in this chapter, have properties which are dominated by those of the fibres.

A key characteristic of the long-fibre composites described in this chapter is their laminated structure, in which the final component consists of a material in which a series of highly anisotropic unidirectional fibre plies are combined together in a particular stacking sequence to provide reinforcement in the appropriate directions to resist the operational loads. This is exemplified in Figure 7.7. This type of construction is commonplace in polymer composite technology and analytical methods for designing such structures have reached a high degree of sophistication and are described fully in standard text books and research papers. Chapter 5 provides an introduction to these techniques and to some of the issues involved in their use for ceramic based materials. A full description of the elastic properties is best carried out by means of a tensor notation which for a unidirectional

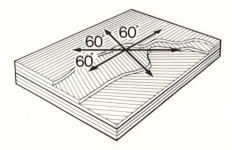

Figure 7.7   Construction of a balanced 60° angle ply laminate.

laminate requires four independent parameters but for an asymmetric, multidirectional laminate, 18 independent terms [30]. In addition, multi-axial failure criteria are required to define the combined conditions of stress under which failure occurs. Again, such failure criteria have been developed for polymer composites.

The highly anisotropic nature of unidirectional, long-fibre composites results in a requirement for totally different test techniques than for homogeneous ceramics and again the techniques developed for advanced polymer composites provide the best starting point for the development of ceramic composite test techniques. An account of some of the issues is given in [31]; a key problem is the adaptation of polymer composite test techniques for high temperature use. Available materials are only beginning to reach the state of optimisation and maturity at which these analytical and test issues are important. Consequently this chapter will be confined to a discussion of mechanical property data which have been generated experimentally in relatively simple tests. For static properties these consist mainly of tensile, flexural, interlaminar shear and compressive strength measurements on unidirectional materials and some measurements on cross-plied (0°, 90°) laminates. These data have been generated principally as part of the process of materials development.

7.4.2   Static mechanical properties

Figure 7.8 shows a typical load versus displacement curve, from a flexural test, for a unidirectional ceramic matrix fibre composite, in this

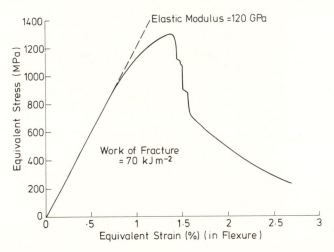

Figure 7.8   Typical load extension behaviour of a SiC fibre reinforced borosilicate glass composite [6].

case a composite consisting of 60 vol% of SiC fibres in a borosilicate glass matrix [6]. At low loads the load-displacement curve is linear and the behaviour is elastic until at some critical load, typically around 0.5–0.7 of the ultimate load, the behaviour becomes non-linear. This non-linearity occurs because the strain to failure of the matrix is lower than that of the fibres and the onset of non-linearity occurs as tensile cracks are produced in the matrix orthogonal to the fibres [32]. Provided the bond between fibres and matrix is not sufficiently strong to permit matrix cracks to propagate through the fibres, the composite can retain its load-bearing capability to higher loads, the matrix crack density increasing until the ultimate strength is achieved. Figure 7.9 shows the regular array of matrix cracks orthogonal to fibres in a unidirectional composite of carbon fibres in a borosilicate glass matrix [32]. At the highest load a catastrophic failure of fibres begins to occur and the load-bearing capability decreases but this can be in a controlled manner, the failure being characterised in a tough composite by much fibre pull-out (Figure 7.10) [2]. Again the toughness of the composite is determined by the fibre–matrix bond strength. Figure 7.11 shows the fibrous nature of the final fracture surface of the SiC/borosilicate glass composite in Figure 7.8. Important parameters in this behaviour sequence are the stress, or strain, at which matrix cracking occurs, the ultimate strength, and the energy absorbed in total failure (the work of fracture) which in the example of Figure 7.11 was $70 \, kJ \, m^{-2}$ [6]. The fibre–matrix bond strength and residual fabrication stresses are important in controlling these properties.

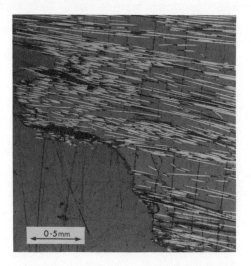

0·5mm

Figure 7.9 Matrix microcracking developed in a carbon fibre reinforced glass composite on loading [2, 32].

Figure 7.10 Fibres bridging the two halves of a fracturing specimen of carbon fibre reinforced glass [2].

Figure 7.11 Fracture surface of a Nicalon SiC fibre/borosilicate glass composite showing the extensive fibre pull-out and matrix cracking: Arrows indicate stressing direction (courtesy of R. W. Davidge).

Figure 7.12 shows the variation of flexural strength with fibre volume fraction for a SiC borosilicate glass composite [6]. A linear dependence of strength is seen as expected from the rule of mixtures. This is largely a consequence of minimising damage to the fibres and producing a low porosity matrix at high fibre volume fractions. Porosity can substantially

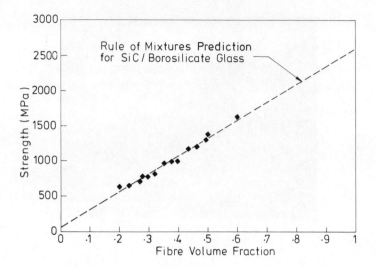

Figure 7.12   Flexural strength of a Nicalon SiC fibre reinforced borosilicate glass composite as a function of fibre volume fraction [6].

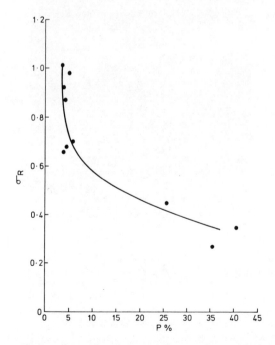

Figure 7.13   The variation of experimental strength, σ_R, expressed as a fraction of the theoretical strength calculated from the rule of mixtures, with the total porosity, expressed as a percentage of the matrix volume, for a carbon fibre reinforced borosilicate glass composite [1, 2].

reduce strength as shown in Figure 7.13 in which the flexural strength of a carbon fibre reinforced glass, normalised by dividing by its theoretical rule of mixtures strength, is plotted against matrix porosity [2].

Interlaminar and transverse strengths of these composites are much lower than the strengths in the direction of the fibres because of the brittleness and low strength of the matrix. Figure 7.14 shows the orientation dependence of strength of a unidirectional borosilicate glass composite [32]. Typical interlaminar shear strengths are given in Table 7.2 and a qualitative account of the effect of bond strength was given earlier.

The advantages of ceramic-matrix composites, particularly of the type described here, over monolithic ceramics include high strength, high toughness and better damage tolerance, and a lower variability of strength. A decreased variability of strength is as important to the designer as the other advantageous properties because it enables him to design components to a higher proportion of their ultimate strength with greater confidence. In ceramic technology, variability of strength is generally characterised through the Weibull approach [33, 34] (see also chapters 4 and 5). The Weibull modulus m of monolithic ceramics and of ceramic fibres is generally low at around 6–8, indicating a high variability in strength. Much higher values have been obtained for ceramic-matrix composites, for example a value of m of 30 for SiC fibre reinforced borosilicate glass [6], indicating a much lower variability in strength. An increase in m from a value less than 10 to a value in the

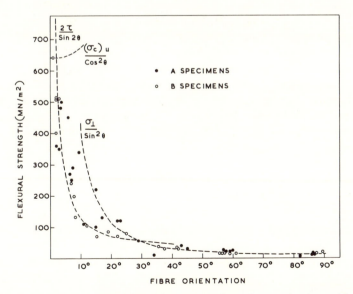

Figure 7.14 The effect of fibre orientation on the flexural strength of a unidirectional carbon fibre reinforced glass [2, 32].

range 20–30 is equivalent, very approximately, to reducing the coefficient of variation in strength from more than 0.12 down to 0.04, a substantial benefit [3].

An as yet unresolved problem in the strength properties of these composites is the relationship between tensile and flexural strength [9]. The measured tensile strengths are always lower than the flexural strengths as shown by typical data in Table 7.4. It is not unusual for the flexural strengths of materials to be higher than their tensile strengths but this can usually be explained through the dependence of strength on stressed volume through a simple Weibull analysis [33, 34]. Such an explanation does not appear to apply to the data in Table 7.4. The ratio between flexural and tensile strength of the carbon–borosilicate composite in Table 7.4 is 2.4. Simple beam theory overestimates the strength of a microcracking composite in flexure because the decrease in modulus as microcracks develop causes a shift in the neutral axis [9, 32]. Allowing for this effect reduced the ratio to 2.2 but the Weibull analysis gave only 1.5. Davies *et al*. concluded [9] that straightforward application of Weibull analysis to relate strength to size may not be appropriate for these composites and suggest that an approach based on a modified fibre bundle failure theory, such as that in [35], may be more appropriate, although it is worth noting that there is much evidence that a Weibull approach can work satisfactorily for polymer-matrix composites. This is an important point which needs clarification before components can be designed with confidence from coupon data.

Another feature of the tensile behaviour of these ceramic composites is that they display a non-constant Poisson ratio [9, 36]. Figure 7.15 for example shows the longitudinal and transverse strains measured as a function of applied load when a undirectional carbon fibre–glass matrix composite was loaded in tension in the fibre direction [36]. The longitudinal strain shows an inflexion at about 350 MPa due to the matrix microcracking described above but there is also an abrupt change in transverse strain from the expected contraction to an expansion. This has been attributed to either 'brooming out' of the fibres as a result of

Table 7.4 Relationship between tensile and flexural strength of two composites (after Davies *et al*. [9])

Composite	Flexural strength (MPa)	Tensile strength (MPa)
Carbon fibre borosilicate glass	1104 ± 77	462 ± 42
Tyranno fibre glass ceramic	1150 ± 150	707 ± 26

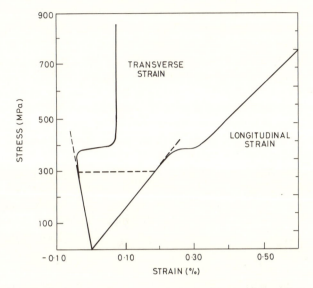

Figure 7.15 Longitudinal and transverse strain as a function of applied stress for a unidirectional carbon fibre reinforced glass composite loaded in tension in the fibre direction, (after Nardone and Prewo [36]).

matrix microcracking [36] or to the occurrence of some axial cracking [9, 37].

The compressive strengths of monolithic ceramics tend to be substantially greater than their tensile strengths but for composites of the type described here this is generally not the case. Compressive failure of unidirectional composites loaded parallel to the fibres is initiated by localised kinking or buckling of the fibres [38] and compressive strengths can be similar to or less than tensile strengths.

7.4.3 *Toughness*

An important property of ceramic-matrix composites is their increased toughness relative to monolithic ceramics. Long-fibre ceramic composites, both those produced by the techniques described in this chapter and those in chapter 8 on CVI composites, can have very high levels of toughness.

Toughness is not a single property like tensile strength, but is better regarded as a class of properties which may be measured and used in design in a number of different ways. The unifying feature underlying different definitions of toughness is that they are all measures of the amount of work that must be done, or energy which is absorbed, in fracturing a component [31, 39, 40]. Confusingly, the term 'fracture toughness' is often used in a narrow and specific sense to mean the critical stress intensity factor, K_{IC}, derived via linear elastic fracture mechanics (LEFM).

Four different measurements of toughness have been used commonly in the development of ceramic-matrix fibre composites. These are the work of fracture (WoF); G_{IC} or critical rate of release of strain energy; a notional, or 'candidate', value of K_{IC} usually referred to as K_Q; and a variety of impact energy measurements. The work of fracture is obtained by measuring the work done in fracturing a specimen in flexure in a controlled manner, such that all the energy is absorbed in the fracturing specimen, and dividing by twice the notional cross-sectional area of the specimen. It is expressed in terms of $J\,m^{-2}$. G_{IC} can be measured in a similar way or else through the relationship between change in stored elastic energy and compliance as a crack propagates,

$$G_{IC} = \frac{P^2}{2} \frac{\partial C}{\partial A}$$

where P is the load and $\partial C/\partial A$ is the change in compliance (C) with crack area (A). K_Q is measured by introducing a crack or notch into a specimen of standard geometry, loading the specimen to failure, and then calculating K_Q through standard LEFM expressions which use the maximum load and the initial crack length. The importance of K_Q, or K_{IC}, for materials for which LEFM is appropriate is that it provides a simple relationship between flaw size (c) and fracture stress (σ), $\sigma \propto K_Q c^{-1/2}$, and contains the controlling material parameters, $K_Q \propto (EG_{IC})^{1/2}$, where E is elastic modulus. Impact energy can be measured by a variety of standard materials testing techniques such as Charpy, Izod or drop-weight impact. For brittle, homogeneous, isotropic materials simple relationships between these parameters can be derived [40] but the situation for highly anisotropic, inhomogeneous, materials such as fibre composites, which can exhibit extensive delocalised damage prior to ultimate failure is complex. Figure 7.16 for example shows the development of damage in a similar polymer composite laminate.

Considerable use has been made of LEFM in the science and technology of monolithic ceramics but it is important to appreciate that LEFM is not, in general, a valid analytical tool for designing with composites of the type described in this chapter, because fundamental assumptions such as localised damage and self-similar crack growth do not apply. Despite this, values of K_Q are frequently quoted for ceramic composites and used to give some feel for the properties of these materials relative to other engineering materials. Typical values of K_Q are 20–30 $MPa\,m^{1/2}$ which should be compared with values of around 5 $MPa\,m^{1/2}$ for typical engineering ceramics and values of around 30 $MPa\,m^{1/2}$ to 200 $MPa\,m^{1/2}$ for tough metals.

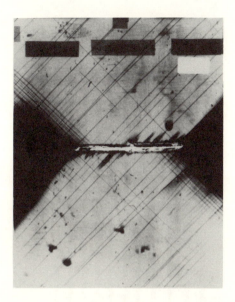

Figure 7.16 X-radiograph of damage developed in an angle-plied laminate of carbon-epoxy, showing the development of interply cracks and delamination (courtesy of R. J. Lee).

Typical values of work of fracture of unidirectional, long-fibre, ceramic-matrix composites lie in the range from a few kJ m^{-2} to ≈ 100 kJ m^{-2} [6, 7, 41], which should be compared with values of a few J m^{-2} for unreinforced glass and glass ceramics to a few hundred J m^{-2} for engineering ceramics. In a flexural test a low work of fracture results in a catastrophic drop in load at fracture, while a high work of fracture results in a very controlled fracture with the load decreasing relatively slowly (cf. a piece of glass and a piece of wood). Indeed the toughness of carbon fibre reinforced borosilicate glass as measured by the work of fracture test is similar to that of teak. The high values of work of fracture result from fibre toughening processes such as those illustrated in Figure 7.17 which describes one possible sequence of events during the propagation of a crack. Again, from polymer composite technology, a good general understanding exists of these processes [42] although the importance of specific models to particular ceramic-matrix systems has only been clarified in one or two cases [7, 8, 41]. An important feature of these models is that a low fibre–matrix bond strength is required for high toughness. More recently the effects of a distribution of fibre strengths on the crack propagation characteristics of ceramic-matrix composites have been investigated theoretically [43, 44].

The interpretation of impact data for composites in terms of engineering design is still at a rudimentary state although considerable advances

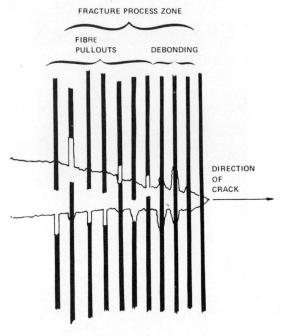

Figure 7.17 An illustration of a possible sequence of events during crack propagation through a fibre composite illustrating some of the energy absorption processes.

have been made in understanding for high performance polymer composites in recent years. Few data currently exist for ceramic-matrix composites of this type. An important engineering case is the situation where a component is subjected to impact while under stress. This has been the subject of detailed research and Figure 7.18 shows data obtained from two different carbon fibre systems, manufactured to have two different bond strengths [45]. This impact map shows that at low imposed stresses the borosilicate glass composite is superior, while at modest stresses ($\approx 10\%$ of ultimate) a transition occurs so that at higher stresses the glass-ceramic composite is superior. This behaviour was reflected in the morphologies of fracture.

7.4.4 *Matrix microcracking*

The phenomenon of matrix microcracking is an extremely important feature of ceramic-matrix composites and has the potential to be a limiting factor in their engineering use by significantly adversely affecting their behaviour under fatigue and at moderate to high temperatures

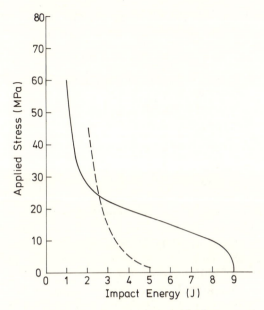

Figure 7.18 An impact map for (–) carbon fibre–LAS glass ceramic and (– –) carbon fibre–borosilicate glass composites [45].

in oxidising atmospheres. These aspects will be described more fully later and the mechanics of matrix microcracking will first be considered.

Matrix microcracking occurs when the failure strain of the matrix is lower than that of the fibre as illustrated in Figure 7.19. The stress at which matrix microcracking might occur, ignoring effects due to thermal

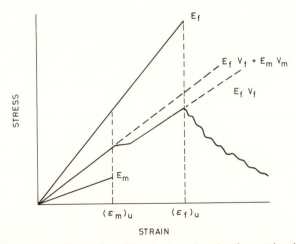

Figure 7.19 Schematic illustration of the simple explanation for matrix microcracking.

mismatch, can be calculated from the simple rule of mixtures which assumes equal strains in fibres (ε_f) and matrix (ε_m)

$$\sigma_c = \sigma_f V_f + \sigma_m V_m \qquad (7.1)$$

where σ_c, σ_f, σ_m are stress in composite, fibre and matrix. Now

$$\varepsilon_m = \frac{\sigma_m}{E_m} = \varepsilon_f = \frac{\sigma_f}{E_f} \qquad (7.2)$$

Therefore

$$\sigma_c' = (\sigma_m)_u \left[1 + V_f \left(\frac{E_f}{E_m} - 1 \right) \right] \qquad (7.3)$$

where $(\sigma_m)_u$ is the strength of the matrix and σ_c' is the stress in the composite at which matrix cracking should occur. It has long been known that this simple formula, in fact, underestimates the stress and strain at which cracking occurs because there are processes which suppress the formation of cracks [2].

Aveston, Cooper and Kelly (ACK) [46] employed an energy balance argument. The formation of a matrix crack is governed by a number of energy terms. These include: the work done by the applied stress in extending the composite, ΔW; the release of strain energy by relaxation of the matrix on either side of the crack, ΔU_m; the resulting increase in strain energy of the fibres in this region, ΔU_f; the frictional energy loss due to relative motion of fibres and matrix near the crack, U_s; and the fracture surface energy of the matrix, $2\gamma_m V_m$. Summing these terms, the necessary condition for spontaneous cracking is

$$\Delta W + \Delta U_m \geqslant \Delta U_f + U_s + 2\gamma_m V_m \qquad (7.4)$$

and it can then be shown that the expected failure strain of the matrix of a composite of high failure strain fibres in a low failure strain matrix is

$$\varepsilon_m = \left(\frac{12\gamma_m \tau E_f V_f^2}{E_c E_m^2 V_m r} \right)^{1/3} \qquad (7.5)$$

where τ is the shear strength of the interface.

More recently Marshall et al. [47] and McCartney [48] have developed another approach to the problem in which they consider the growth of a crack opposed by tractive forces imposed by the crack-bridging effect of the fibres. This approach provides a greater insight into the processes involved in microcrack initiation and growth than the ACK theory and predicts a flaw size/failure relationship of the form shown in Figure 7.20. For flaws greater than a critical size the stress at which matrix microcracking occurs is independent of flaw size and approximates to the ACK value (equation (7.5)). At lower flaw sizes the stress increases

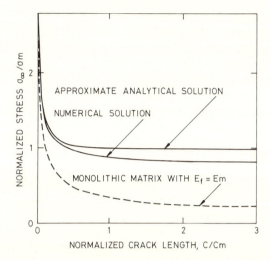

Figure 7.20 The variation of matrix crack propagation stress with crack size (after Marshall *et al*. [47]).

as flaw size decreases. The critical flaw size approximates to a few fibre diameters and it is believed that for present, practical systems, fabrication and machining flaws are of a size such that the matrix microcracking stress is independent of flaw size and well described by equation (7.5). The matrix microcracking stress can thus be most easily increased by increasing the fibre matrix bond strength and the fibre volume fraction. It might also be increased, less easily, by decreasing the fibre diameter and by increasing the ratio E_f/E_m.

A contribution to the suppression of matrix microcracking can also be obtained through thermal expansion mismatch between fibres and matrix by placing the matrix in compression through choosing $\alpha_f > \alpha_m$, as explained earlier, although for isotropic fibres this will result in a contraction of the fibre away from the matrix and a potential decrease in fibre–matrix bond strength.

For reasons described later it is known that matrix microcracking can cause early fatigue failure and initiate early failure at high temperatures. By considerations such as the above it might be possible to produce high strength systems in which microcracking is suppressed to stresses and strains close to the ultimate strength with consequent improvements in dynamic fatigue, static fatigue and high temperature lifetime. In developing such systems it is important to optimise the bond strength carefully as too high a bond strength will result in a brittle, low toughness composite.

An important feature of the Marshall, Cox and Evans theory is its prediction of a flaw size dependence of matrix microcracking stress for flaw sizes smaller than the critical size. There is interest in protecting these composites from environmental attack by means of coatings. If composites are eventually developed in which the intrinsic flaw size is less than the critical flaw size then the coating thickness may become the dominant flaw size and thin coatings may be better than thick coatings. Again, clearly, there will be a need for optimisation of a number of conflicting requirements.

7.4.5 *Temperature effects and fatigue*

Matrix microcracking might impose a significant limitation on the engineering use of ceramic composites through its effect on fatigue life and on strength at temperature.

A major limitation to the use of carbon fibre reinforced glasses and glass ceramics is their oxidation in air at temperatures in excess of about 350–400 °C [2, 49]. In non-oxidising atmospheres strength is retained to temperatures at which creep of the matrix causes loss of strength, or reaction between the carbon fibre and the constituents of the glass occur, in principle well in excess of 1000 °C for glass-ceramic systems.

Silicon carbide fibres yield composites which can operate to much higher temperatures in oxidising atmospheres but problems occur even for these [50]. Figure 7.21 from work by Prewo [50] shows the flexural strength of a SiC fibre reinforced lithium–alumino–silicate system, denoted by Prewo as LAS-I. This, relatively low strength system retained its strength to a temperature in excess of 1000 °C in air. Figure 7.22 from the same work shows the behaviour of a much stronger SiC/LAS system, LAS-II. This composite displayed a marked decrease in strength at 700–800 °C with a change to a more brittle, planar fracture at the tensile stressed surface. The system LAS-I, because of its

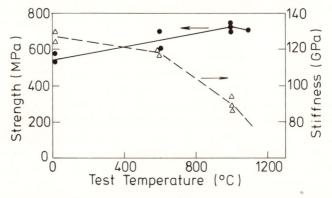

Figure 7.21 The flexural strength (solid line) and stiffness (broken line) of a SiC fibre–glass ceramic composite (LAS-I) tested in air (after Prewo) [50]).

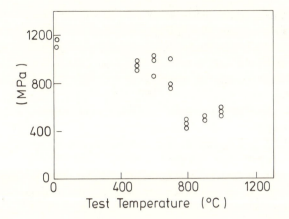

Figure 7.22 The flexural strength of a SiC fibre–glass ceramic composite (LAS-II) tested in air (after Prewo [50]).

relatively low strength, did not display extensive matrix microcracking while the higher strength system, LAS-II, did. It has been established that the rapid loss of strength of LAS-II at 700–800 °C was due to a mechanism involving atmospheric attack of the carbon rich interface between fibres and matrix promoted by the matrix microcracks which permitted easy access of oxygen to the interface [27]. Electron microscopy studies have provided some insight into the detailed mechanisms involved. A similar effect has been recorded in SiC–SiC composites again associated with matrix cracking and it is significant that although rapid loss of strength occurs at 800 °C with that system, it did not occur at 1400 °C [51]. Matrix microcracking can thus play a significant role in controlling the strength at temperature of even SiC composites.

Another feature of matrix microcracking is that it can control fatigue behaviour. In polymer-matrix composites matrix microcracking can induce fatigue failure and by analogy this might be expected to occur in ceramic-matrix systems. At present there are few published fatigue data at high cycles and the indications are not conclusive for CMC fatigued in non-aggressive environments, although similar temperature effects to those described above for static strength occur for SiC-fibre composites fatigued in air.

Early studies on carbon fibre reinforced glasses investigated the effect of fatigue at room temperature into the matrix microcracked region to fatigue lives in excess of 10^7 cycles [2]. No significant change in residual strength was observed although there was a change in the morphology of fracture. The density and penetration of matrix cracking increased on cycling and fatigued specimens, tested in flexure after cycling, failed in a more catastrophic way with a reduced work of fracture.

More detailed studies have been carried out by Prewo on SiC fibre reinforced glass-ceramic systems although only to 10^5 cycles [52]. In his experiments, which also include static fatigue tests, Prewo showed that there was no significant degradation up to 10^5 cycles when materials were cycled into matrix microcracking at room temperature, but that severe degradation occurred at 600 °C and 900 °C, associated with the similar effect observed on static strength due to oxidative attack of the carbon-rich interface.

Holmes et al. [17] have studied the fatigue of hot-pressed SiC fibre reinforced Si_3N_4 at high temperatures and have shown that there is a rapid reduction in fatigue strength on cycling into the matrix micro-cracked region. At $\approx 10^6$ cycles the fatigue strength had decreased to the proportional limit, although there is no evidence that this is a fatigue limit.

It thus appears important to either develop systems in which micro-cracking is suppressed to levels close to the ultimate strength or to develop fibre–matrix interfaces which are resistant to oxidative attack.

In situations in which strength at temperature is not affected by microcracking the maximum temperature which might be achieved by ceramic fibre reinforced glass ceramics are determined either by the maximum temperature capability of the matrix or of the fibres. Table 7.1 from Prewo and Brennan [4] shows the maximum use temperature of some glass ceramics and suggests that temperatures as high as 1500–1700 °C might be possible. However currently available oxide and non-oxide fibres are limited to substantially lower temperatures. Short term tests of strength at temperature indicate that the strength of available oxide fibres are halved at around 1100 °C and of available SiC-type fibres at around 1400 °C [53]. These temperatures arc significantly reduced under creep conditions as shown in chapter 2. The current maximum temperature capability of ceramic-fibre composites for long term use under stress in oxidising atmospheres is probably around 1200–1300 °C. More refractory fibres will be necessary before the higher temperature matrices in Table 7.1 become useful. For inert atmospheres, however, carbon fibres which retain their strength to temperatures in excess of 2000 °C in non-oxidising atmospheres, might be a satisfactory reinforcement.

References

1. Sambell, R. A. J., Phillips, D. C. and Bowen, D. H. (1974) The technology of carbon fibre reinforced glasses and ceramics, in: *Carbon Fibres, Their Place in Modern Technology, Proc. Int. Conf.*, Plastic Institute, London.
2. Phillips, D. C. (1983) in: *Handbook of Composites, Vol. 4, Fabrication of Compo-*

sites, eds. Kelly, A. and Mileiko, S. T., Elsevier, Amsterdam, Chapter 7.
3. Phillips, D. C. (1989) Fibre reinforced ceramics, in: *Concise Encyclopaedia of Composite Materials*, ed. Kelly, A., Pergamon Press, Oxford.
4. Prewo, K. M. and Brennan, J. J. (1989), in: *Reference Book for Composites Technology, Vol. 1*, ed. Lee, S. M., Technomic, Lancaster, PA, Chapter 6.
5. Rice, R. W. and Lewis III, D. (1989) in: *Reference Book for Composites Technology, Vol. 1*, ed. Lee, S. M., Technomic, Lancaster, PA, Chapter 7.
6. Dawson, D. M., Preston, R. F. and Purser, A. (1987) *Ceram. Eng. Sci. Proc.* 8, 815–821.
7. Phillips, D. C. (1974) *J. Mater. Sci.* 9, 1847–1854.
8. Aveston, J. (1971) Strength and toughness in fibre reinforced ceramics, in: *Proc. Conf. Properties of Fibre Composites,* IPC Science and Technology Press Ltd., Guildford.
9. Davies, J. J. R., Preston, R. F., Lee R. J. and Walls, K. N. (1990) Fabrication and mechanical properties of improved fibre reinforced ceramics, in: *New Materials and Their Applications*, Proceedings of a Conference held at the University of Warwick, Institute of Physics, London.
10. Hegeler, H. and Brükner, R. (1989) *J. Mater. Sci.* 24, 1191–1194.
11. Davidge, R. W. Harwell laboratory, private communication.
12. Bowen, D. H. (1989) Manufacturing methods for composites: an overview, in: *Concise Encyclopaedia of Composite Materials,* ed. Kelly, A., Pergamon Press, Oxford.
13. Hyde, A. R. (1988) *GEC J. Res.* 6, 44–49.
14. Layden, G. K. and Prewo, K. M. (1987) Pultrusion of glass and glass-ceramic matrix composites, US Patent No. 4, 664, 731.
15. Sakomoto, H., Kodama, H. and Hiyoshi, T. (1987) *J. Ceram. Soc. Jpn. Int. Ed.* 95, 817–822.
16. Kodama, H., Sakamoto, H. and Hiyoshi, T. (1989) *J. Amer. Ceram. Soc.* 72, 551–558.
17. Holmes, J. W., Kotil, T. and Foulds, W. T. (1989) High temperature fatigue of SiC fibre reinforced Si_3N_4 ceramic composites, in: *Symp. on High Temperature Composites, Proc. Am. Soc. for Composites*, Technomic, Lancaster, PA.
18. Lundberg, R. (1989) Fibre reinforced ceramic composites, Doctorate Thesis, Department of Inorganic Chemistry, Chalmers University of Technology, Gothenberg, Sweden.
19. Lundberg, R., Pompe, R., Carlsson, R. and Goursat, P., *Composites Sci. Technol.* 37, 165–176.
20. Haggerty, J. S. (1989) *Mater. Sci. Eng.* A107, 117–125.
21. Materials and Processing Report, Vol. 2, No. 3, June 1987.
22. Materials and Processing Report, Vol. 3, No. 3, June 1988.
23. Aghajanian, M. K., Macmillan, N. H., Kennedy, C. R., Luszcz, S. J. and Roy, R. (1989) *J. Mater. Sci.* 24, 658–670.
24. For example see: Michalske, T. A. and Hellman, J. R. (1988) *J. Am. Ceram. Soc.* 71, 725–731.
25. Mah, T., Mendiratta, M. G., Katz, A. P. and Mazdiyasni, K. S. (1987) *Ceram. Bull.* 66, 304–308.
26. Brennan, J. J. (1986) Interfacial characterisation of glass and glass-ceramic matrix/Nicalon SiC fibre composites, in: *Proc. Conf. on Tailoring Multiphase and Composite Ceramics*, eds. Tessler, R. T., Messing, G. L., Pantano, C. G. and Newnham, R. E., Plenum, New York.
27. Cooper, R. F. and Chyung, K. (1987) *J. Mater. Sci.* 22, 3148–3160.
28. Murty, V. S. R. and Lewis, M. H. (1989) *J. Mater. Sci. Lett.* 8, 571–572.
29. Bleay, S. M. and Scott, V. D. (1991) Microstructure property relationship in Pyrex glass composites reinforced with Nicalon fibres, *J. Mater. Sci.*, in press.
30. For a fuller account see standard texts, e.g. Jones, R. M. (1975) *Mechanics of Composite Materials*, Scripta Book Co.
31. Phillips, D. C. and Davidge, R. W. (1986) *Br. Ceram. Trans. J.* 85, 123–130.
32. Phillips, D. C., Sambell, R. A. J. and Bowen, D. H. (1972) *J. Mater. Sci.* 7, 1454–1464.

33. Davidge, R. W. (1979) *Mechanical Behaviour of Ceramics,* Cambridge University Press, Cambridge.
34. Davidge, R. W. (1987) Engineering performance prediction for ceramics, in: *Fracture of Non-Metallic Materials,* eds. Herrmann, K. P. and Larsson, L. H., EEC, Brussels, pp. 95–115.
35. Davidge, R. W. and Briggs, A. (1989) *J. Mater. Sci.* 24, 2815–2819.
36. Nardone, V. C. and Prewo, K. M. (1988) *J. Mater. Sci.* 23, 168–180.
37. Harris, B. University of Bath, private communication.
38. Lankford, J. (1987) *Composites* 18, 145–152.
39. Davidge, R. W. and Phillips, D. C. (1972) *J. Mater. Sci.* 7, 1308.
40. Phillips, D. C. and Harris, B. (1977) The strength, toughness and fatigue properties of polymer composites, in: *Polymer Engineering Composites*, ed. Richardson, M. O. W., Elsevier Applied Science, London.
41. Phillips, D. C. (1972) *J. Mater. Sci.* 7, 1175–1191.
42. Beaumont, P. W. R. (1989) Toughness of fibrous composites, in: *Concise Encyclopaedia of Composite Materials*, ed. Kelly, A., Pergamon Press, Oxford.
43. Sutcu, M. (1988) *J. Mater. Sci.* 23, 928–933.
44. Thouless, M. D. and Evans, A. G. (1988) *Acta Metall.* 36, 517–522.
45. Phillips, D. C., Park, N. and Lee, R. J. (1990) *Composites Sci. Technol.* 37, 249–266.
46. Aveston, J., Cooper, G. A. and Kelly, A. (1971) Single and multiple fracture, in: *Proc. Conf. on the Properties of Fibre Composites*, IPC Science and Technology Press, Guildford.
47. Marshall, D. B., Cox, B. N. and Evans, A. G. (1985) *Acta Metall.* 33, 2013–2021.
48. McCartney, L. N. (1987) *Proc. R. Soc. London Ser.* A 409, 329–350.
49. Prewo, K. M. and Batt, J. A. (1988) *J. Mater. Sci.* 23, 523–527.
50. Prewo, K. M. (1986) *J. Mater. Sci.* 21, 3590–3600.
51. Frety, N. and Boussuge, M. (1990) *Composites Sci. Technol.* 37, 177–190.
52. Prewo, K. M. (1987) *J. Mater. Sci.* 22, 2695–2701.
53. Pysher, D. J., Goretta, K. C., Hodder Jr., R. S. and Tressler, R. E. (1989) *J. Am. Ceram. Soc.* 72, 284–288.
54. Holloway, D. G. (1973) *The Phyical Properties of Glass*, Wykeham Publications, London.

8 CVI composites

R. NASLAIN

8.1 Introduction

The principle of the CVI processing of ceramic-matrix composites (CMCs) has been given in chapter 3. It consists of depositing a ceramic matrix from a gaseous precursor in the pore network of a preform made of ceramic fibres [1, 2]. Starting from a fibre preform with an open porosity ranging usually from 40–80%, the CVI process yields CMCs with a residual porosity of the order of 10–15%. An important advantage of the CVI process lies in the fact that it is performed at relatively low temperatures (typically 900–1100 °C). As a result, the brittle ceramic fibres remain undamaged during processing even those of limited thermal stability (e.g. the Nicalon or transition alumina fibres). Furthermore, in CVI, the starting material is usually a preform which can be of complex shape thus allowing the preparation of near net shape parts. Finally, no external pressure is applied on the preform during CVI densification, a feature which permits the utilisation of continuous fibres which can be pre-arranged in beneficial orientations that remain undisturbed during the process.

On the other hand, CVI can be applied only to ceramic matrices: (i) for which gaseous precursors are readily available (i.e. covalent or ionocovalent ceramics) and (ii) of simple chemical compositions. Finally, CVI yields CMCs which do exhibit some residual open porosity, a feature which can be a drawback for applications where gas or liquid tightness is required.

The feasibility of the CVI process has already been established for a number of ceramic matrices including carbon, carbides (SiC, B_4C, TiC), nitrides (BN, Si_3N_4) and oxides (Al_2O_3, ZrO_2). It has been shown that CVI composites, when properly processed, are tough ceramic materials whose applications at high temperatures are limited more by the weakness of available ceramic fibres and the nature of the interphase than by the intrinsic properties of the CVI matrices. As far as the author knows, only *SiC matrix CVI composites* are currently produced on an industrial scale, mainly for applications in the aeronautic and space fields.

8.2 SiC matrix CVI composites

8.2.1 *Processing*

8.2.1.1 The fibre preform. In the CVI processing of CMCs, the starting material is a *preform* which is made from ceramic fibres and has already the shape of the final part. Depending on the nature of each particular part, the fibres used to make the preform are either short fibres (i.e. chopped fibres or whiskers) or continuous fibres. Taking into account the chemical compatibility requirements, three different kinds of ceramic fibres can be used to reinforce a silicon carbide matrix: (i) *SiC-based yarn fibres* resulting from the pyrolysis of a polycarbosilane (PCS) precursor (e.g. the Nicalon fiber), (ii) *SiC whiskers* or (iii) *carbon fibres*. The use of ex-PCS fibres is limited to 1200–1400 °C due to crystallisation and grain growth phenomena. The high-temperature properties of SiC whiskers are still unknown and their availability as textile products is not yet achieved. Finally, carbon fibres which are probably the best reinforcement for a SiC-matrix from a thermal stability and mechanical point of view, are very sensitive to oxidation beyond 500–600 °C. Thus, the choice of the reinforcement should be made on the basis of the environmental conditions for each part.

Short-fibre preforms can be obtained as shown in Figure 8.1a. In a first step, the short fibres are suspended in a liquid containing a binder, for example polycarbosilane. The slurry is then filtered to form a disc (or any other simple shape). Finally, the disc is heated under an inert atmosphere. During this last step, the polycarbosilane binder is thermally decomposed into a SiC(0) phase binding the fibres together. Such preforms are characterised by: (i) a rather low fibre volume fraction (15–25%), (ii) a fibre orientation which can be regarded as orthotropic (i.e. random in a plane) and (iii) a poor ability to be densified by CVI inasmuch as the initial open pores become prematurely sealed by the ceramic deposit [3].

Continuous-fibre preforms are preferred when high fibre volume fractions and a given degree of anisotropy are required. As a matter of fact, 1D*-*preforms* (i.e. preforms in which the fibres are orientated in only one direction) are rarely used in industry due to: (i) the weakness of the resulting CMCs in a transverse direction and (ii) their poor ability to be densified by CVI. On the other hand, very high fibre volume fractions could be achieved theoretically with 1D fibre architectures. Therefore, the preforms actually used are *nD-preforms*, *n* being usually equal to two or three. *2D-preforms* are made from a stack of ceramic fabrics maintained close together either with a ceramic tool (*dry*

*Note that D is usually taken to mean direction rather than dimension in continuous-fibre, multidirectional preforms.

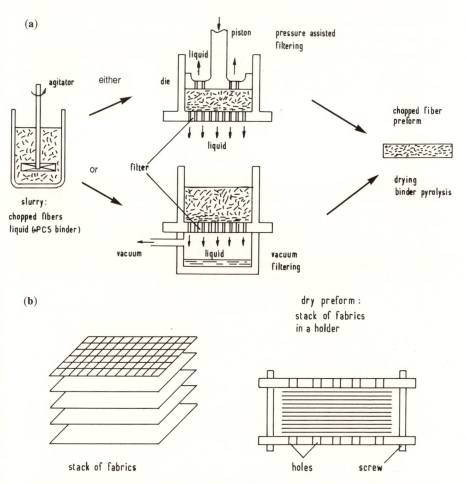

Figure 8.1 Fibrous preforms processed from: (a) a slurry of chopped fibres containing a polycarbosilane binder (consolidated preforms) and (b) a stack of ceramic fabrics maintained pressed with a holder (dry preforms).

preforms) or with a small amount of binder (*consolidated fabrics*) (Figure 8.1b). In the first case, the ceramic tool is retained during the first part of the CVI process until the fabrics are bonded enough by the ceramic deposit, then it is removed. In the second case, the binder, which can be of the same or different nature than the matrix, is usually applied according to a liquid phase procedure similar to that described above for short-fibre preforms. These two kinds of preforms lead to CMCs which may have different properties since, in the former, the CVI matrix is deposited directly on the reinforcement (i.e. in the pores of small diameter present within the tows as well as in those of much larger diameters located between adjacent fabrics) whereas, in the latter, it is deposited on the binder (i.e. only in the inter-fabric layer large pores, the small pores within the tows being already filled with the binder itself). The open overall porosity of cloth preforms are typically of the order of 40–50%. Finally, *nD-preforms* (with $n > 2$) are obtained using specific manual or automatic weaving techniques, the *3D-preforms* where the fibres are orientated along three orthogonal directions being of particular interest [3–5].

In order to be easily densified by CVI, a preform should exhibit a pore *spectrum* with a high fraction of large size pores open and interconnected. As discussed in the next sections, these pores whose diameters range from 10 to 500 μm allow the in-depth diffusion of the CVI reactants and products. The pores of smaller diameters are much more difficult to fill by CVI. Therefore the pore spectrum appears to be a very important preform parameter in the processing of CMCs by CVI [6].

As already discussed in chapter 4, obtaining CMCs with high toughness and work of fracture requires a *compliant interphase* between the fibres and the matrix. This interphase is usually made of a thin film of carbon or boron nitride acting as a mechanical fuse to protect the brittle fibres against the notch effect arising from the matrix microcracking. There are several ways to form this interphase during the processing of CMCs . It can be deposited on the fibres prior to the preparation of the preforms (C-coated SiC-reinforcements are currently available though only with CVD-filaments not as yarn fibres). It can also be formed during the preparation of the preforms. As an example, in consolidated preforms, the binder may contain the interphase material and, in dry preforms, the first ceramic material deposited by CVI may be the interphase material. Although the thickness of the interphase necessary to achieve a non-brittle mechanical behaviour has to be optimised for each particular CMC, it is typically of the order of 0.1–1 μm [1, 7–9].

8.2.1.2 The isothermal/isobaric process (ICVI). The principle of the CVI-processing of CMCs from porous fibre preforms has been given in chapter 3. CVI is a processing technique which is directly derived from chemical vapour deposition (CVD). In both techniques, a solid (here silicon carbide) is formed at medium temperatures from the same gaseous precursor (e.g. a mixture of a chlorosilane and hydrogen, for SiC). The main difference between them lies in the fact that in CVI the deposition of the solid phase occurs within the pore network of the substrate (in depth deposition or infiltration) whereas in CVD it is limited to the external surface of the substrate (surface deposition or coating). Furthermore, in order to fill as completely as possible the open porosity of the substrate, which is indeed the objective in CVI processing, the pore entrances of the substrate should remain open up to the end of the densification process. In order to fulfil this requirement, CVI has to be preformed under conditions which are rather different from those commonly used in classical CVD, as discussed below [2, 10–12].

In isothermal/isobaric CVI (ICVI), the preform is set within a hot wall deposition chamber and its temperature is assumed to be uniform (i.e. there is no temperature gradient in the substrate). Furthermore, in ICVI the mass transfers (of both gaseous reactants and products) along the pore network occur only by diffusion (i.e. no pressure gradient is applied to the substrate). Under such conditions, the deposition process can be rate-limited by two different mechanisms, as shown schematically in Figure 8.2 for a single straight cylindrical pore open at both ends: (i) the *mass transfers of the reactants* and products in the gas phase along the pore and across a stagnant boundary layer surrounding the substrate and (ii) the *kinetics of the surface phenomena* including the chemical reaction itself. In order to obtain a deposit of rather uniform thickness along the pore, the kinetics of the surface reaction should be slow with respect to the reactant mass transfer by diffusion. Under such conditions, the reactant molecules diffuse in the pore far from the pore entrance before being adsorbed on the pore wall and reacting. If this condition is not fulfilled, that is if the kinetics of the chemical reaction is fast with respect to the reactant mass transfer, the reactant molecules are consumed immediately when arriving near the hot external surface

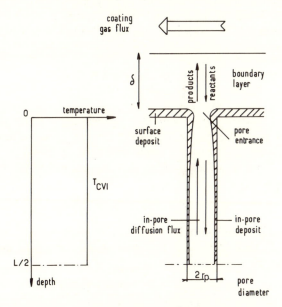

Figure 8.2 Chemical vapour infiltration of a straight cylindrical pore according to the ICVI process (after [14]).

of the substrate and the deposit is limited to the vicinity of the pore entrance. As a result, the pore entrance will be rapidly sealed by the deposit and the densification of a porous preform will be obviously impossible [2, 12, 13].

The competition between the two rate-limiting steps can be discussed on the basis of a *CVD dimensionless number X* taking into account the surface reaction kinetic constant k_s, an effective diffusion coefficient D_e and a geometrical parameter which can be the pore diameter $2r_p$ in the simple case of the straight *cylindrical pore* shown in Figure 8.2 and a *first order reaction* (e.g. that leading to the deposition of SiC from CH_3SiCl_3 in the presence of hydrogen) [13, 14]:

$$CH_3SiCl_{3(g)} \xrightarrow{H_2} SiC_{(s)} + 3HCl_{(g)} \qquad (8.1)$$

$$X = \frac{k_s r_p}{D_e} \qquad (8.2)$$

(other expressions for the dimensionless number, involving a different geometrical parameter, have also been used) [6, 10]. The kinetics constant k_s is known to obey an Arrhenius law with respect to temperature:

$$k_s = k_0 \exp(-Q/RT) \qquad (8.3)$$

where k_0 is a pre-exponential constant, Q the activation energy and R the gas constant. Generally speaking, the effective diffusion ceofficient should include both the *Fick diffusion* coefficient D_F and the *Knudsen diffusion* coefficient D_K, which can be combined according to the classical equation

$$\frac{1}{D_e} = \frac{1}{D_F} + \frac{1}{D_K} \qquad (8.4)$$

For pores of rather *large diameters* (e.g. $2r_p > 100 \ \mu m$), Knudsen diffusion can be neglected with respect to the Fick diffusion and D_e which is equal to D_F to a first approximation, is known to depend on both T and P for a given gaseous species, according to the following equation:

$$D_f = D_0 T^m P^{-1} \qquad (8.5)$$

where D_0 is a constant and $1.5 < m < 2$. In such a case, by combining equations (8.3) and (8.5) with equation (8.2), the CVD dimensionless number X can be written as

$$X = \frac{k_0 r_p}{D_0} \times P \times \frac{\exp(-Q/RT)}{T^m} \qquad (8.6)$$

The value of X can be used to discuss the thickness homogeneity of the deposit along a pore of given radius r_p. When X is small, that is when the surface reaction kinetics is slow with respect to reactant mass transfer by diffusion that is to say the CVI experiments *performed at low T and P*, the deposit is expected to be rather uniform along the pore. In contrast, when X is high, that is for experiments performed at high T and P, deposition will be limited to the vicinity of the pore entrance.

For pores of *small diameters* (e.g. $2r_p < 10 \ \mu m$), the Knudsen diff-usion can no longer be neglected (it may be the main mass transfer mechanism for pores of very small diameters at a low enough pressure). Under such conditions, D_e must be calculated according to equation (8.4), D_K being given according to the gas kinetic theory by the following equation:

$$D_K = \frac{2r_p}{3} \left[\frac{8RT}{M} \right]^{1/2} \qquad (8.7)$$

for a gaseous species of molar mass M in a pore of radius r_p. It is worthy of note that D_K does not depend on total pressure P. Although the resulting expression for the CVD dimensionless number is more complex than that given by (8.6), the main conclusions drawn for the effect of both P and T on the deposit thickness uniformity in the pore remain valid. However, below a pressure threshold (where Knudsen diffusion becomes the main mechanism for the reactant mass transfer), it is expected that total pressure P is no longer an effective CVI parameter.

The *modelling of the deposit thickness profile* along a cylindrical pore, has been done recently by different authors for the deposition of SiC from CH_3SiCl_3/H_2 [14–16]. All the models, although they differ from a mathematical viewpoint, result qualitatively in the same conclusions as far as the effects of temperature, total pressure and pore radius are concerned. As an example, the results of calculations due to Naslain *et al.* [14] utilizing a model derived by Van den Brekel/Rossignol *et al.* [13, 17], are shown in Figures 8.3 and 8.4. The deposit thickness profiles along the pore (assumed to be open at both ends and of length $L = 10$ mm) were computed according to an iterative procedure stopped when the deposit thickness at the pore entrance (i.e. at $z = 0$) is equal to the initial pore radius r_p, that is when the pore becomes sealed at both ends by the deposit.

The result of the calculations confirms that the thickness homogeneity is excellent when temperature and pressure are low (e.g. $T = 800$–$900 \ °C$; $P = 2$–20 kPa) at least when the pore diameter is large enough ($2r_p = 100 \ \mu m$), that is for low values of X. In contrast, raising T or/and P yields deposits non-uniform in thickness (the deposit being much thicker near the pore entrance). This feature is still more evident

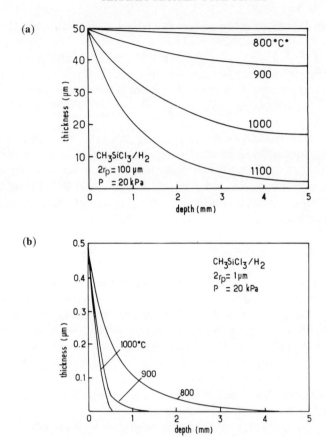

Figure 8.3 SiC-deposition thickness profiles along a straight cylindrical pore calculated, for different temperatures, according to the Van den Breckel/Rossignol model [14]: (a) pore of large diameter ($2r_p = 100 \ \mu$m); (b) pore of small diameter ($2r_p = 1 \ \mu$m).

for pores of small diameters ($2r_p = 1 \ \mu$m). In fact, almost no deposition occurs beyond $L/10$ from the pore entrance when $T = 1000\,°C$ and $P = 20\,kPa$. Furthermore, lowering temperature to 800 °C only slightly improves the deposit profile whereas lowering total pressure to 2 kPa has no effect at all, a result suggesting that, under such conditions, Knudsen diffusion may be the only mass transfer mechanism of the reactants along the pore. These qualitative conclusions, drawn from simple theoretical considerations, are in agreement with the experimental data reported independently by Christin *et al.* for 2D-C–C consolidated fibre preforms [12, 18] and by Fitzer *et al.* for both cylindrical model pores and porous monolithic substrates [6, 11].

In CVI, the porous fibre preforms (dry or consolidated preforms) are

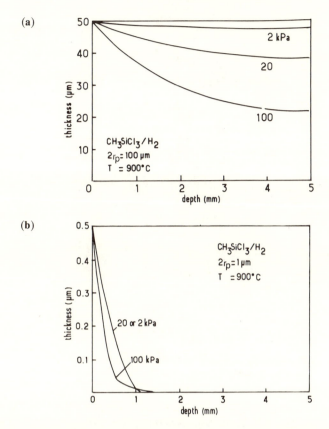

(a)

(b)

Figure 8.4 SiC-deposition thickness profiles along a straight cylindrical pore calculated, for different pressures, according to the Van den Breckel/Rossignol model [14]: (a) pore of large diameter ($2r_p = 100 \ \mu$m); (b) pore of small diameter ($2r_p = 1 \ \mu$m).

set inside a hot-wall isothermal infiltration chamber, as shown schematically in Figure 8.5. The infiltration chamber is fed with the gaseous precursor (a mixture of CH_3SiCl_3 and hydrogen, for SiC) flowing under a reduced pressure. The optimal values of T and P depend on both the initial pore spectrum of the preforms and the degree of pore filling by SiC (the pore becoming smaller and smaller in diameter as densification proceeds). For given T and P values, the densification rate (i.e. the mass of SiC matrix deposited per unit of time) *decreases progressively versus time* [19]. It thus becomes very low when the pores are almost totally filled and a full densification might require several hundreds of hours. Practically, it may be more appropriate to perform the infiltration at rather high T and P values (to increase the densification rate) and to keep them constant during the whole process. However, under such

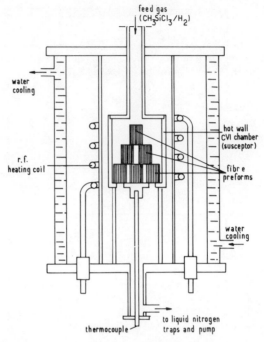

Figure 8.5 Hot-wall deposition chamber (schematic) used for the processing of SiC-ceramic-matrix composites according to the ICVI technique.

conditions, the pores have to be reopened several times by surface machining. Finally, for economic reasons, the infiltration process is often stopped when the residual porosity has been reduced to about 10–15%.

The nature of the ceramic matrix deposited from a CH_3SiCl_3–H_2 gaseous precursor depends mainly on the value of the molar ratio $\alpha = [H_2]/[CH_3SiCl_3]$ (and to a lesser extent on temperature and pressure). It consists of: (i) *pure silicon carbide* (cubic β modification with small amounts of hexagonal polytypes) for medium α values (e.g. $1 < \alpha < 10$), (ii) *SiC + C* mixtures for low α values and (iii) *SiC + Si* mixtures for high α values (Figure 8.6) [12, 18].

The *main drawback* of the ICVI process lies in the low densification rate related to the specific T and P conditions under which infiltration has to be performed when the reactant mass transfer in the preforms is only by diffusion. However, this drawback is compensated by the fact that: (i) a large number of preforms are treated simultaneously and (ii) the process can be applied to any preform shape (either simple or complex) provided it is not too thick. As far as we know, it is the only CVI process presently used for the production of CMCs at an industrial level [20].

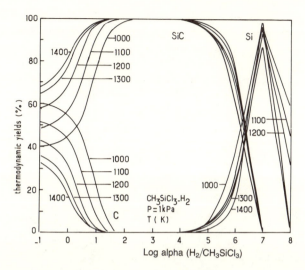

Figure 8.6 Nature of the solid deposited at equilibrium from a mixture of CH₃SiCl₃ and H₂, as a function of both temperature and initial composition (dilution ratio of the source species $\alpha = H_2/CH_3SiCl_3$), derived from thermodynamic calculations. SiC is deposited as a pure single phase for medium α-values whereas it is co-deposited with either carbon or silicon for low and high α-values, respectively [14].

8.2.1.3 The temperature/pressure gradient process (FCVI). Temperature or/and pressure gradients can be applied to a porous preform during infiltration in order to: (i) avoid an early pore sealing by SiC (temperature gradient) and/or (ii) increase the infiltration rate (pressure and temperature gradients).

It has been shown recently, on the basis of mathematical modelling, that the infiltration of a cylindrical pore by SiC (deposited from CH₃SiCl₃/H₂) could be significantly improved when a *temperature gradient*, even as low as 5%, is applied along the pore, the *lower temperature being at the pore mouth*. Under such conditions, in-depth densification is favoured (see equation (8.3)) and the deposition front moves from the inner sections of the pore towards its entrance. As a result, a complete filling of the pore by SiC is achieved (for $T = 800–1000\,°C$ at pore entrance) without any early pore sealing [16].

In order to increase the infiltration rate of SiC, Caputo *et al.* have proposed that the diffusion mass transfer be replaced by *forced convection* (forced CVI or FCVI). As schematically shown in Figure 8.7, such a mass transfer by forced convection is obtained by applying a *pressure gradient* along the pores. Furthermore, a *temperature gradient* can be superimposed, as discussed above, on the pressure gradient (in the opposite direction). Practically, the reactants are injected through one face of the preform under a high pressure P_1 (with $P_1 = 100–200\,kPa$)

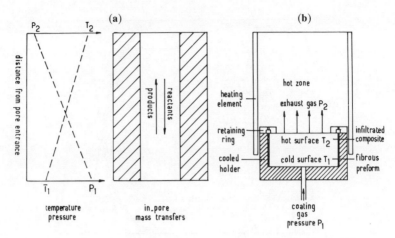

Figure 8.7   The FCVI process: (a) temperature and pressure gradients applied along a straight cylindrical pore, (b) experimental set-up (after [14, 22]).

at a low temperature T_1 and the exhaust gases are pumped (pressure P_2) through the opposite face at a higher temperature T_2 (with $T_2 = 1100$–$1200\,°C$). In order to generate the pressure and temperature gradients, each preform must be set within a specific ceramic tool (retaining ring, heater and cooled holder). Due to the high value of T_2 and to the fast reactant mass transfer, the upper hot face of the preform is rapidly coated by SiC and becomes impermeable. Therefore, the reactant flow which is initially axial soon becomes radial (the exhaust gas escaping from the preform laterally). Finally, as the matrix is deposited the thermal conductivity of the infiltrated portions of the preform increases and the deposition zone progressively moves from the top of the preform towards the bottom and the circumference [3, 21, 22].

A *one-dimensional model* of the FCVI (axial flow step) by SiC of a porous preform made of random short fibres, has been proposed by Starr [23]. As shown schematically in Figure 8.8, when $T_1 = 1000\,°C$ and $T_2 = 1200\,°C$, the deposition rate is almost uniform along the preform axis whereas when $T_1 = T_2 = 1200\,°C$ (isothermal CVI), deposition occurs preferentially near the face through which the reactants are injected in the preform, as discussed in section 8.2.1.2, due to a rapid depletion of the gas phase at this high deposition temperature (equation (8.3)). It was concluded that fine tuning of the infiltration process could be achieved by controlling both T_1 and the gas flow rate. A more complete *two-dimensional model*, simulating the radial flow step and the effect of the different temperature parameters (temperatures of the gas inlet and outlet, preform faces and preform holder), has been worked out by Tai and Chou [24].

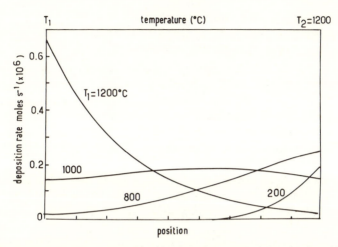

Figure 8.8 A one-dimensional modelling of the FCVI process giving the deposition rates of SiC (from CH_3SiCl_3/H_2) at different locations in the preform, for various temperatures. When $T_1 = 1000\,°C$ and $T_2 = 1200\,°C$ (see Figure 8.7), an almost uniform infiltration occurs within the whole preform (after [14, 22]).

One of the main advantages of the FCVI process lies in the fact that the time necessary to reach a given degree of densification of a given preform by SiC is *reduced by one order of magnitude* with respect to that characterizing the ICVI process. Furthermore, the FCVI process is well suited to the densification of thick wall fibre preforms [25]. On the other hand and as already mentioned, FCVI cannot be applied readily to the simultaneous treatment of a large number of preforms particularly when they are of complex and different shapes (due to tool requirement).

8.2.2   Properties

8.2.2.1   Microstructure and anisotropy.   In view of the potential of laminates in many applications and the ability of 2D-fibre preforms to be densified by CVI, most data on the properties of SiC-matrix CVI composites, available from the literature, deal with 2D-carbon fibre/SiC matrix or 2D-SiC (Nicalon) fibre/SiC matrix composites. In the following sections, the discussion will be limited to these two materials referred to as 2D-C/SiC and 2D-SiC/SiC.

Since 2D-composites are *anisotropic* materials, the orientation of the reinforcement in the specimens as well as the direction along which a load is applied and/or a property is measured, should be clearly indicated. By convention, *directions 1 and 2* are in the plane of the cloth (the warp and woof weaving directions) whereas *direction 3* is perpendicular to the cloth (stacking direction of the fabrics in the preform), as shown in Figure 8.9 [26–29].

When observed in cross-sections, a 2D-SiC matrix composite obtained according to the CVI process exhibits a relatively *coarse microstructure*, with respect to sintered polycrystalline SiC ceramics, the infiltrated tows being usually clearly apparent. As already mentioned in section 8.2.1, the residual porosity is of the order of 10–15% with small pores within the tows and large pores between the tows (Figure 8.10) [3, 18, 30].

8.2.2.2 Mechanical behaviour at room temperature. When *loaded in tension* along directions 1 or 2, a 2D-SiC–SiC composite (Nicalon fibres; $V_f \approx 0.45$; $V_p \approx 10\%$) processed under optimised conditions, exhibits a

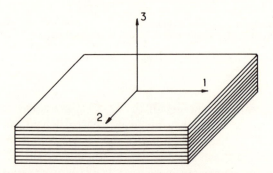

Figure 8.9 Orientation of the axes in a 2D-ceramic-matrix composite processed from fibre cloths according to the CVI techniques (after [26]). 1 = warp direction; 2 = woof direction.

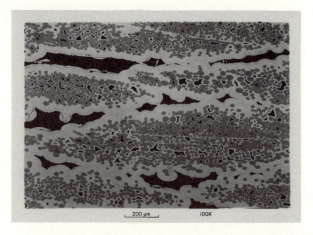

Figure 8.10 Microstructure of a 2D-SiC–SiC composite processed according to the FCVI technique from a CH_3SiCl_3/H_2 precursor: large pores are located between the fibre tows and small pores within the infiltrated tows (from [3]).

non-linear behaviour, with a strain to failure usually of the order of 0.3% [20, 29] but which can be as high as 0.6–0.8% in some cases [26, 31, 32]. Such features are unusual for a SiC-based ceramic material but have been already reported for other brittle-matrix composites, as discussed in chapters 4 and 7.

As shown in Figure 8.11, the stress–strain curve, recorded in tension on unnotched specimens, exhibits two domains [29, 33]. In domain I, that is below a stress level of about 100 MPa (which can be assumed to be the yield stress), the material behaves linear elastically with a Young's modulus $E_1 = 200$–210 GPa. However, on the basis of acoustic emission, some damaging phenomena already occur from around 60 MPa. In domain II, that is for $\sigma > 100$ MPa, the stress–strain curve is no longer linear due to the extension of damaging phenomena typical of CMCs in which the fibres are not too strongly bonded to the matrix through a compliant interphase (here a layer of pyrocarbon). These damaging phenomena, which have been discussed in chapter 4, are mainly matrix microcracking, fibre–matrix debonding and fibre–matrix sliding. Moreover, when the specimen is unloaded from a point of the curve located in domain II and then reloaded, the residual strain (at $\sigma = 0$) and the area of the unloading/loading loop are very small. Therefore, 2D-SiC–SiC composites can be described, at least in a first approximation, as *quasi-linear elastic materials* whose stiffness decreases progressively through the development of damage mechanisms. The non-linear behaviour of the material is less apparent in tests performed

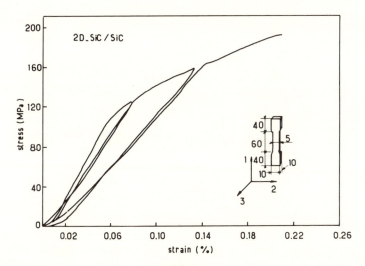

Figure 8.11 Stress–strain curve in tension and at room temperature of a 2D-SiC/SiC composite prepared from a 2D-SiC (ex-PCS) preform according to the ICVI process (after [33]).

in flexure (three or four point bending tests) and almost not apparent at all in compression [29, 33].

The *toughness of ceramic materials* is usually characterised, from tests performed on notched specimens, on the basis of the concepts of the linear elastic fracture mechanics (LEFM), for example the critical stress intensity factor K_c and the critical strain energy release G_c. However, these concepts initially introduced for brittle homogeneous isotropic materials and then extended to metals (under conditions where the plastic deformation zone is of small volume and limited to the vicinity of the crack tip) may not be applicable straightforwardly to CMCs and may not lead to intrinsic materials parameters, due to the specific features mentioned above (i.e. coarse microstructure with respect to specimen size, significant residual porosity, non-linear mechanical behaviour, large damage zone associated with a macrocrack propagation). This important field is presently the subject of active research [27–29, 33, 34]. Therefore, the data reported in literature must be considered with care. In fact, they are usually presented more as a simple way to compare homologous materials or to optimise processing conditions than as intrinsic material characteristics.

Bernhart *et al.* have used two different parameters, referred to as K_r and K_{Ir}, to characterise the apparent toughness in rupture, of SiC-matrix composites from tests performed on SENB or CT specimens. The former was used for materials in which the initial crack is deflected initially from its plane and the latter when the crack does propagate in its initial plane (depending on the fibre architecture and orientation of

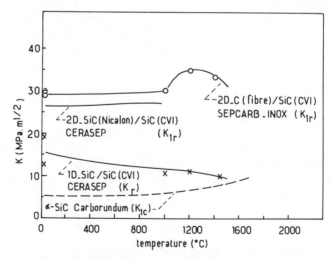

Figure 8.12   Variations of the apparent toughness of SiC-matrix composites processed by CVI, as a function of test temperature, according to [26].

the notch). As shown in Figure 8.12, apparent toughness K_{Ir} values as high as *25–30 MPa m$^{1/2}$* are given for both 2D-C–SiC and 2D-SiC (Nicalon)–SiC composites, at room temperature. These values, with the limitations formulated above, are much higher than the K_{Ic} values reported for monolithic SiC-based ceramics, that is 4–5 MPa m$^{1/2}$, and similar to that mentioned for 2024 or 6061 (T6 conditions) light alloys often used for aerospace applications, that is 22–30 MPa m$^{1/2}$. However, they remain rather far from the K_{Ic} of tough titanium alloys or steels (i.e. 50–90 MPa m$^{1/2}$ [26]).

In a recent study, Bouquet *et al.* have shown that 2D-SiC (Nicalon)–SiC composites deviated only slightly from linear elastic behaviour, as already mentioned, and concluded that the LEFM concepts, for example the crack *growth resistance curve* (or *R*-curve), could be used to obtain at least relative data on toughness which may be useful for the optimisation of the material [29]. As shown in Figure 8.13a, the *R*-curve gives the strain energy release rate when a crack, initiated at the tip of a notch of length a_0, propagates under the effect of a load *P* applied to the specimen in, for example a three point bending configuration. It is established on the basis of the following equation:

$$G_I = \frac{1}{2} \frac{P^2}{B} \frac{dC}{da} \tag{8.8}$$

where *C* is the actual compliance of the damaged material, *B* the specimen thickness and *a* the crack length. A compliance calibration curve $C = f(a/W)$ allows the determination of an effective crack length from a compliance measurement. It was observed, for a specimen loaded in direction (1, 2), that two systems of crack propagation take place alternately. The first is related to the fracture of a SiC-fibre tow (acting as a crack arrestor) whereas the second is the propagation of the crack within the SiC CVI matrix deposited in the pores between the tows. It appears from Figure 8.13a that the general shape of the *R*-curves is not very dependent on the notch length. Although a critical value G_{Ic} could not be defined as an intrinsic material parameter, as already mentioned, it was suggested that the average value of $\Delta G_I/\Delta a$ be used for the purpose of material comparison based on toughness considerations.

R-curves were also drawn from slow crack growth tests performed on notched compact tension samples (20 mm square) of 2D-SiC–SiC composites, by Bernhart *et al.* [26, 31]. The notch was opened slowly, with a periodic release of the load to measure the residual compliance of the damaged material and thence the crack length (Figure 8.13b). When performed on sintered SiC ceramics, used as a standard, the test yields a flat curve which corresponds to a low and constant value of crack

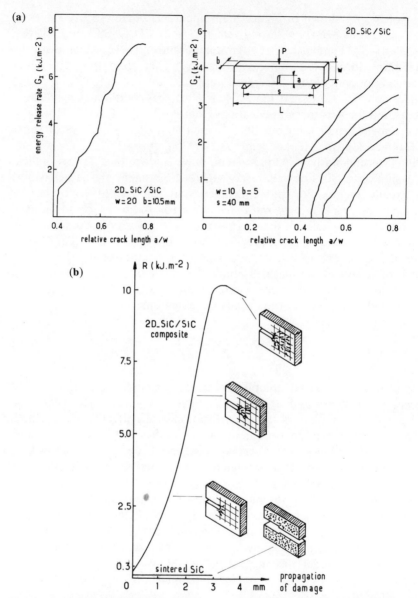

Figure 8.13 Resistance to crack propagation of 2D-SiC/SiC composites processed according to the ICVI technique: (a) *R*-curves corresponding to notches of different lengths, (b) *R*-curves of composites and sintered ceramics (after [29, 31]).

propagation energy, a typical feature of the catastrophic failure of brittle monolithic ceramics. For the composite, on the other hand, resistance to crack propagation increases in a dramatic manner as the crack propagates, up to energy values which are two orders of magnitude higher than those measured for the sintered SiC standard. The high fracture energy is thought to be connected to multiple branching of the crack and to the extension of a damaged zone in the material. The *R*-curve concept was also used by Gomina *et al.* to characterise the mechanical behaviour of homologous 2D-C–SiC composites [28, 34].

The *statistical character* of the failure strength was studied in experiments performed in 4 point bending, on the basis of Weibull $\ln \ln (1/P_s) = f(\ln \sigma)$ plots where P_s is the probability of survival of the sample loaded at a stress level σ (see chapter 4). For 2D-SiC–SiC composites loaded in direction (1, 3), the value of the Weibull modulus m was observed to be of the order of 14 despite microstructural defects (e.g. residual pores), which is a high value for SiC ceramics [26].

From the above discussion, it appears that C–SiC and SiC–SiC composites obtained according to the CVI process (e.g. from 2D-fibre preforms) are *tough and reliable materials* compared to the corresponding unreinforced ceramics, despite the fact that they exhibit a rather high residual porosity. These properties are observed when a weak interphase of pyrocarbon or boron nitride, allowing fibre pull-out, is present. When this is not the case, the material is brittle, as discussed in chapter 4. The presence of pyrocarbon- and BN-interphases has been demonstrated using microanalytical techniques such as TEM, AES and SIMS [8, 35].

8.2.2.3 Thermal properties. The *coefficient of thermal expansion* (CTE) of a 2D-SiC/SiC composite, calculated from thermal expansion measurements performed in directions 1 or 2 between 25 and 1000 °C, that is $\alpha_1 \approx 4 \times 10^{-6}$ °C^{-1}, is similar to that of a sintered silicon carbide ceramic [32]. The thermal expansion of 2D-C–C preforms and that of the related 2D-C–C/SiC composites have been measured for temperatures ranging between room temperature and 2200 °C, in both the 1 (or 2) and 3 directions [36]. The starting 2D-C–C consolidated preforms are strongly anisotropic but their thermal expansion anisotropy progressively decreases as the porosity is filled with SiC. As a result, the difference between the α_3 and α_1 values is smaller and smaller as the volume fraction of SiC deposited in the porosity increases, as shown in Figure 8.14a. In fact, 2D-C–C/SiC derived from strongly anisotropic 2D-C–C preforms appears to be almost isotropic, as far as CTE is concerned, when totally densified with SiC.

The thermal diffusivity of 2D-SiC/SiC composites has been measured from room temperature up to 1500 °C, in both directions 1 (or 2) and 3 [31]. It is anisotropic at room temperature but the difference between a_3 and a_1 decreases as temperature increases, as shown in Figure 8.14b. That of 2D-C–C/SiC composites has been measured in direction 3 by Naslain *et al.* [36].

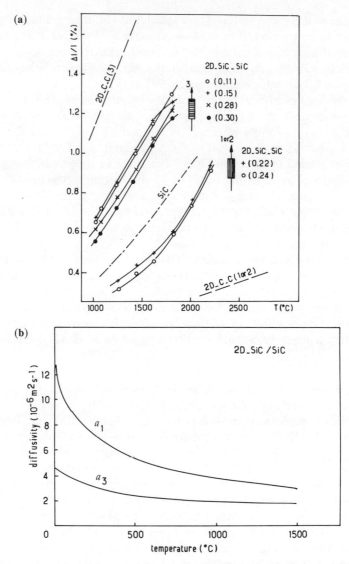

Figure 8.14 Thermal properties of 2D-ceramic/ceramic composites measured along directions 1 or 2 (i.e. parallel to the fabric layers) and 3 (perpendicular to the fabric layers): (a) thermal expansion of 2D-C–C/SiC composites, derived from a 2D-C–C preform (ex-PAN carbon fabrics stacking consolidated with pyrocarbon), the SiC volume fractions are given in parentheses with $V_{SiC} + V_p = V_{po} = 0.30$. SiC and 2D-C–C fully densified with pyrocarbon are also included [36]. (b) thermal diffusivity of 2D-SiC–SiC composites (derived from 2D-SiC (Nicalon) fabrics (after [31])).

8.2.2.4 Effect of the environment. The strength of both 2D-SiC/SiC and 2D-C/SiC composites has been measured at *high temperatures*, under an atmosphere of argon/hydrogen (4 point bending test). As

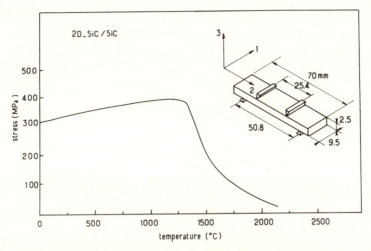

Figure 8.15 High-temperature bending failure stress of 2D-SiC/SiC composites measured under an atmosphere of Ar/H$_2$ (after [31]).

shown in Figure 8.15, the failure stress first rises progressively as temperature is increased, up to 1300 °C for 2D-SiC/SiC and presumably well above this value for 2D-C/SiC composites. Then, it decreases rather sharply, at least for the 2D-SiC/SiC composites, for $1300 < T < 1500$ °C, due to the strength loss of the SiC (Nicalon) fibres known to occur within this temperature range. However, it is worthy of note that the failure strength of the composite is still 50% of its room temperature value at 1500 °C and that load bearing capacity of the composite is still significant at 2000 °C [31, 32]. In the same manner, the fracture toughness (K_{Ir}) has been reported to increase slowly as temperature is raised for both 2D-SiC/SiC and 2D-C/SiC composites [26, 31, 32]. The occurrence of a maximum for 2D-C/SiC composites, at about 1200 °C, comparable to that reported for SiO$_2$–glass-ceramic matrix composites, remains to be confirmed.

Silicon carbide is known for its *good oxidation resistance* at high temperatures related to the formation of a protective layer of silica. In fact, SiC/SiC composites have been designed to sustain long exposures in oxygen-containing atmospheres at high temperatures; therefore, their oxidation resistance is expected to be good. The effect of ageing treatments in air on the residual flexural strength measured at room temperature of 2D-SiC/(Nicalon)–SiC composites was studied by Lamicq *et al.* [31] and more recently by Cavalier *et al.* [20]. No strength loss occurs after ageing treatments performed below 1100 °C (even when the specimens are maintained at this temperature for 500 h). In contrast as shown in Figure 8.16, different changes are observed in the stress –strain curve for treatments performed in air (duration 20 h) above

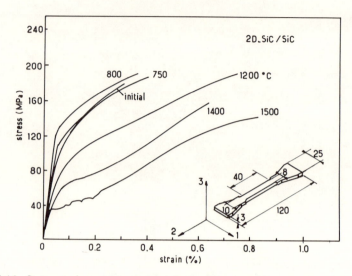

Figure 8.16 Stress–strain curves, recorded in tension at ambient temperature, of 2D-SiC/SiC (CVI) composites after an ageing treatment in air at temperatures ranging from 750 to 1500 °C (after [28]).

1200 °C, namely: (i) a decrease in the stiffness (domain I) and failure stress of the material, as well as (ii) an increase in the failure strain. As already mentioned, these features are partly the consequence of the change occurring in the microstructure of the fibres within this temperature range. Finally, under conditions of thermal cycling (between 100 and 700 °C in air), a slight increase in the residual strength (measured at room temperature) is observed as the number of cycles is increased [31].

Although both the fibres and matrix have intrinsically a good oxidation resistance in a SiC/SiC composite, this is not necessarily the case for the interphase, particularly when it is made of carbon. In such a case, an impermeable protective coating could be applied to the composite to avoid a rapid oxidation of the carbon interphase that will result in a tough to brittle transition in the mechanical behaviour similar to that reported for the SiO_2-base glass or glass-ceramic matrix CMCs [37, 38]. The use of a boron nitride interphase, a material whose oxidation resistance is better than that of carbon, might improve the overall behaviour of 2D-SiC/SiC composites in air at medium temperatures and for long exposures.

2D-C/SiC composites are theoretically more attractive than 2D-SiC/SiC composites for applications at very high temperatures, because carbon fibres are intrinsically more resistant to structural and mechanical degradation than SiC (Nicalon) fibres at $T > 1400$ °C. However, they are unfortunately more sensitive to oxidation. The oxidation resistance of 2D-C–C/SiC composites, obtained by CVI from 2D-C–C consolidated preforms, has been studied by Naslain *et al.* [12, 36]. For a given temperature and oxygen partial pressure, the loss of weight due to the oxidation of carbon decreases as the volume fraction of SiC deposited in the preform porosity increases. As shown in Figure 8.17, it becomes very low for almost fully densified preforms, a feature establishing that the SiC-matrix does provide some protection against oxidation to the carbon components of the material, at least for short exposures. On the other hand, long exposures in air at high temperatures would require an external protective coating, as already discussed for 2D-SiC/SiC composites with a carbon interphase.

Both 2D-C/SiC and 2D-SiC/SiC composites are expected to have a better *thermal shock resistance* than sintered SiC-ceramics, on the basis of their non-linear stress–strain curves and lower stiffnesses. The thermal shock resistance has been assessed by Dauchier *et al.* from experiments where samples were first heated (up to 1000 °C) and then

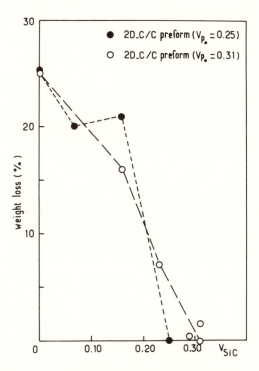

Figure 8.17 Weight loss of 2D-C–C/SiC composites as a function of the SiC volume fraction (with $V_{SiC} + V_p = V_{po}$). Test conditions: $T = 1500$ °C; duration: 7 h; air flow rate: 2 l/h. The weight loss becomes almost nil when $V_{SiC} = V_{po}$, i.e. when $V_p = 0$ (after [12, 36]).

quenched in water at 20 °C [32]. From the residual strength it was shown that neither 2D-C/SiC nor 2D-SiC/SiC composites were significantly affected by thermal shock even at 1000 °C.

So far as the author is aware no data have yet been published on the creep and mechanical fatigue resistances of SiC-matrix CVI composites.

8.3 Other non-oxide matrix CVI composites

The CVI process, which has been described in detail in section 8.2 for the SiC matrix, can be extended to other refractory non-oxide matrices (e.g. carbides, nitrides or silicides) when the following conditions are fulfilled: (i) availability of a suitable gaseous precursor, (ii) deposition performed under reaction kinetic rate-limited regime and (iii) matrix chemically compatible with the fibres (usually carbon fibres) at deposition temperature (to avoid any significant fibre/matrix reaction that will result in a fibre strength loss). The feasibility of the process has been demonstrated on the laboratory scale, for refractory carbide (i.e. TiC and B_4C) and nitride (i.e. BN) matrices. The process has also been used, apparently with less success, for Si_3N_4.

8.3.1 *Processing*

Titanium carbide is usually deposited from a $TiCl_4$–CH_4–H_2 precursor. Thermodynamic considerations show that it is obtained as a pure phase, that is free of elemental carbon, according to the following overall reaction:

$$TiCl_{4(g)} + CH_{4(g)} \xrightarrow{\;H_2\;} TiC_{(s)} + 4HCl_{(g)} \qquad (8.9)$$

only when hydrogen is added to the $TiCl_4$ and CH_4 source species [12, 13]. For low temperature and total pressure conditions, compatible with pore infiltration: (i) the P_{H_2}/P_{CH_4} ratio in the gas precursor must be higher than 10, (ii) a large amount of $TiCl_4$ and $TiCl_3$ gaseous by-products are formed and (iii) as a result, the thermodynamic yield in TiC is low (i.e. less then 30%). Experimental kinetic data show that the deposition process of TiC is rate-limited by surface reaction kinetics below about 950 °C for $P = 10$ kPa. Optimised conditions for a laboratory-scale hot-wall infiltration chamber and 2D-C–C consolidated preforms $(25 < V_p < 60\%)$, have been reported to be $T = 950$ °C, $P = 3$ kPa and $\alpha \approx 10$ [13].

Boron carbide is deposited from a BCl_3–CH_4–H_2 precursor, according to the following overall equation:

$$4BCl_{3(g)} + CH_{4(g)} + 4H_{2(g)} \rightarrow B_4C_{(s)} + 12HCl_{(g)} \qquad (8.10)$$

As shown in Figure 8.18, the deposition process is rate-limited by surface reaction kinetics at temperatures lower than 900–950 °C (the limit depending slightly on total pressure). For a laboratory-scale apparatus and 2D-C–C consolidated preforms ($V_{p_0} = 40\%$), optimised CVI parameters were reported to be close to those mentioned for TiC, that is $T = 950$ °C; $P = 2.6$ kPa; flow rates: $BCl_3 = 20$; $CH_4 = 5$; and $H_2 = 20$ cm^3 min^{-1} [39].

The *kinetics of densification* by ICVI of 2D-C–C consolidated preforms by TiC or B_4C obey similar laws [13, 39]. The densification rate M_p/t (i.e. the mass of ceramic matrix deposited per unit of time in the pore network) is rather high at the beginning of the process but it progressively decreases versus time as the pores become smaller and smaller in diameter. In contrast, the deposition rate on the external surface of the preform $\Delta M_s/\Delta t$ remains constant as shown in Figure 8.19 for boron carbide (inset). Furthermore, it has been established by Rossignol *et al.* (for TiC) and Hannache *et al.* (for B_4C) that the kinetics of densification obey the following linear relation:

$$\ln\left[(M_o - M_p)/M_o\right] = -\lambda t \qquad (8.11)$$

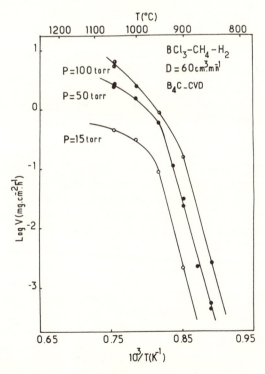

Figure 8.18 Arrhenius plot of the deposition rate of B_4C as a function of reciprocal temperature for various total pressures (D = gas flow rate) (after [39]).

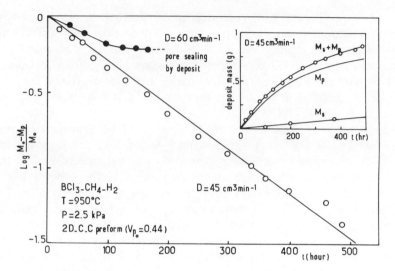

Figure 8.19　Infiltration kinetics of a 2D-C–C preform (made of a stack of carbon fabrics consolidated by CVI with pyrocarbon) by B_4C (deposited from BCl_3–CH_4–H_2) (after [2, 39]).

where M_p is the mass of ceramic deposited in the pore network at time t, M_o is that corresponding to a total filling of the initial preform porosity, V_{po}, and λ is a constant which depends on T and V_{po} [13, 39]. As shown in Figure 8.19, this relation may remain valid almost up to the end of the densification process, under optimised CVI conditions. A deviation from this linear relationship indicates inappropriate infiltration conditions (i.e. an early pore sealing by the deposit). It is worthy of note, as already mentioned in section 8.2 for SiC, that totally filling the available porosity of the preform by the matrix in ICVI may require several hundreds of hours if: (i) no intermediate surface machining of the preform is done and (ii) the CVI parameters are kept constant.

Boron nitride is usually deposited from a gaseous precursor made of a mixture of ammonia and boron halide BX_3 (with $X = F$ or Cl), according to the following overall equation:

$$BX_{3(g)} + NH_{3(g)} \rightarrow BN_{(s)} + 3HX_{(g)} \qquad (8.12)$$

BF_3 is often preferred to BCl_3 for economical reasons and because it results in more crystalline deposits [40, 41]. However, from thermodynamic considerations it appears that: (i) a large percentage of BF_3 remains unreacted at equilibrium under CVI conditions (i.e. the yield in BN is low and typically of the order of 15%), (ii) gaseous by-products are formed, for example boron subfluorides and species arising from an attack of the carbon preforms at the very beginning of deposition, such as HCN or CH_4 [41]. Finally, BF_3 and NH_3 form addition compounds at room temperature which remain solid up to about 300 °C. As a result, the reactants must flow separately in the CVI unit and they are mixed only in the hot infiltration chamber, in order to avoid a sealing of the pipes by solid ammonium fluoroborates. Optimised ICVI conditions, for 2D-C–C preforms, have been given by Hannache *et al.* [41, 42].

Finally, Fitzer mentioned that the ICVI of porous substrates (e.g. RB-Si_3N_4) by *silicon nitride*, deposited from mixtures of ammonia and $SiCl_4$, is much more difficult than that by silicon carbide due to: (i) a much higher deposition rate and (ii) a tendency for gas phase nucleation to take place. Under the best conditions (i.e. $T = 1000$ °C), only 20% of the initial porosity was actually filled with Si_3N_4 with a preferred deposition near the pore entrances resulting in an early pore sealing [6, 43]. Better results have been reported by Caputo *et al.* concerning the densification by Si_3N_4 of preforms made of chopped SiC (Nicalon) fibres according to the FCVI process and utilising a precursor made of a mixture of $SiCl_4$, NH_3 and H_2. However, the final density, that is 50–65% of theoretical density, is significantly lower than that (70–80%) mentioned for the same preforms densified by SiC [3].

8.3.2 *Properties*

The properties of the CVI composites with a ceramic matrix consisting of TiC, B_4C, BN or Si_3N_4 have not been studied as extensively as those of the related SiC-matrix composites. The only available data must be considered as preliminary data, reported mainly for establishing the feasibility of the CVI process.

From a *microstructural* standpoint, these composites have many features in common with the SiC-matrix composites prepared from the same kind of preform. They do contain some residual porosity, the overall value and pore spectrum of which depend on the way the CVI has been performed. An example of the evolution of the pore spectrum is given in Figure 8.20 for B_4C-matrix composites which shows that the majority of the large pores ($2r_p > 1$ μm) are filled by ICVI [2, 39].

A comparative study has been made of the mechanical behaviour of 2D-C–C composites with a variety of matrices by Rossignol *et al.* [44]. Because of the limited volume of material, use was made of compression testing. Tests were made as a function of the volume fraction of ceramic deposited, V_I and thus as a function of residual porosity, $V_p (V_p = V_{po} - V_I)$. The stress–strain curves generally exhibit two regions: (i) a region of linear elastic behaviour and (ii) a region of non-linear behaviour extending up to failure and corresponding to

(a)

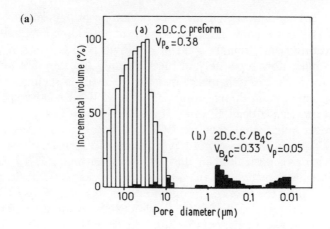

(b)

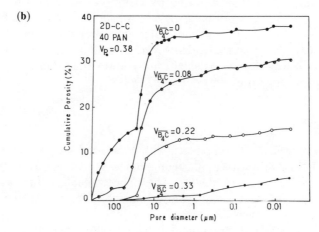

Figure 8.20 Variations of the pore spectrum during the infiltration of a 2D-C–C preform by B_4C: (a) pore spectra of the initial preform and almost fully densified composite, (b) pore spectra of the composite at different stages of the infiltration process (after [2, 39]).

different damaging phenomena. The relative extension of the two regions depends on the value of V_I, the nature of the matrix and the loading direction. The variations of the elastic moduli and failure (or damaging) strengths σ^R as a function of $1 - V_p$ are shown in Figure 8.21. As could be expected, the composites are stiffer and stronger when V_p tends to zero (i.e. $1 - V_p$ tends to 1). The ranking was as follows: 2D-C–C/BN; 2D-C–C/SiC + C; 2D-C–C/TiC; 2D-C–C/SiC and 2D-C–C/B_4C similar to that existing among the unreinforced matrices. This ranking is also shown in Figure 8.22. Models were proposed to take into account the variations of both E and σ^R with the degree of densification of the preforms.

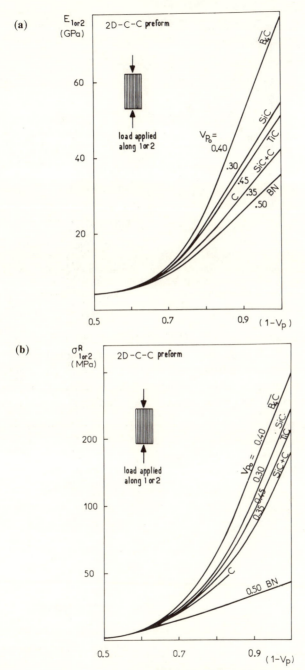

Figure 8.21 Variations as a function of $1 - V_p$ (V_p: residual porosity) of the Young's moduli (a) and failure stress (b) in compression loading (directions 1 or 2) of 2D ceramic-matrix composites processed from 2D-C-C preforms (ex-PAN fibres) according to the ICVI technique (after [14]).

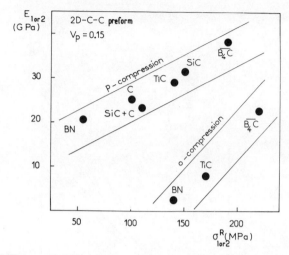

Figure 8.22 Stiffness and failure stress of various ceramic-matrix composites processed from 2D-C–C preforms according to the ICVI technique (with a residual porosity of the same order). The nature of the ceramic matrix is mentioned near each data point (after [44]).

The *oxidation resistance* of 2D-C–C/TiC; 2D-C–C/B$_4$C and 2D-C–C/BN composites was studied as a function of residual porosity V_p, temperature and oxygen partial pressure [42, 45–47]. An example of oxidation kinetic curves is shown in Figure 8.23 for 2D-C–C/TiC composites (residual porosity less than about 5%) tested in air at temperatures

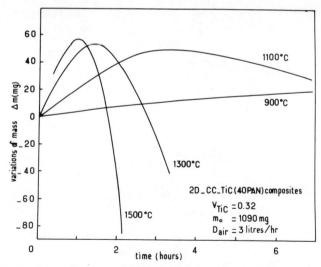

Figure 8.23 Variations of mass of a 2D-C–C/TiC composite during ageing treatments performed at various temperatures under a flow of 3 l of air per hour (atm. pressure) (after [46]).

ranging from 900 to 1500 °C. The shape of the curves can be analysed on the basis of two different oxidation reactions: that of the TiC-matrix leading to titanium and carbon oxides with a weight increase of the sample (stage 1) and that of the carbon preform giving rise to an evolution of carbon oxides with a weight loss (stage II). The oxidation of TiC is predominant at the beginning of the test. Later on due to the non-protective character of the titanium oxide layer and the occurrence of a microcrack network in the TiC-matrix, oxygen diffuses towards the carbon preform whose oxidation becomes the prevailing mechanism. The transition from stage I to stage II occurs at a time which depends on temperature, oxygen partial pressure and residual porosity [45, 46].

The occurrence of two different oxidation mechanisms has also been observed for 2D-C–C/BN composites (residual porosity of about 10%) maintained isothermally in a flow of air at temperatures ranging from 500 to 1100 °C (Figure 8.24). As temperature is raised there is first a weight loss, for $T < 900$ °C, due to the oxidation of the carbon preform by oxygen diffusing through the pore network, then for $T > 900$ °C, a weight increase due to that of the BN-matrix. At a given temperature, the weight loss depends again on the residual porosity and partial

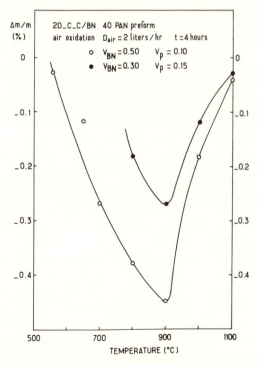

Figure 8.24 Weight loss of a 2D-C–C/BN composite during ageing treatments performed at various temperatures under a flow of 2 l of air per hour (atm. pressure) (after [47]).

pressure of oxygen [42]. It thus appears that the two important parameters controlling the oxidation resistance of a 2D-C–C/ceramic composite are: (i) the residual porosity and (ii) the protective character of the solid (or liquid) oxides resulting from the oxidation of the ceramic matrix.

8.4   Oxide-matrix CVI composites

The CVI process has been further extended to refractory oxide matrices, that is to *alumina* and *zirconia*. These oxides have high melting points and interesting physical, mechanical and chemical properties. Gaseous precursors (e.g. MCl_x -H_2-CO_2 mixtures, with $x = 3$ for $M = Al$ and $x = 4$ for $M = Zr$) are either readily available on the market or easily prepared *in situ* from the metals. However, an important limitation is the lack of suitable ceramic fibres. On the one hand, carbon fibres have the required mechanical properties at high temperatures but they react with most (if not all) CVI refractory oxide matrices and they are very sensitive to oxidation, as already discussed. On the other, oxide fibres for example alumina based fibres, undergo a dramatic strength loss at moderate temperatures due to either crystallisation, phase transformation and/or grain growth phenomena.

8.4.1   *CVI processing of oxide-matrix composites*

Alumina and zirconia matrices can be infiltrated in a porous fibre preform by *in situ* hydrolysis of the corresponding chlorides $AlCl_3$ and $ZrCl_4$. An example of an experimental arrangement is shown schematically in Figure 8.25. The chloride is obtained by chlorination of the metal at about 300 °C while water is conveniently formed in the infiltration chamber itself heated at 900–1000 °C, by oxidation of hydrogen with carbon dioxide. The depositions of alumina or zirconia, from $AlCl_3$–H_2–CO_2 and $ZrCl_4$–H_2–CO_2 precursors, thus correspond to the following chemical equations:

$$Zr + 2Cl_2 \rightarrow ZrCl_4 \qquad (8.13)$$
or
$$2Al + 3Cl_2 \rightarrow 2AlCl_3 \qquad (8.13')$$

$$H_2 + CO_2 \rightleftharpoons H_2O + CO \qquad (8.14)$$

$$ZrCl_4 + 2H_2O \rightarrow ZrO_2 + 4HCl \qquad (8.15)$$
or
$$2AlCl_3 + 3H_2O \rightarrow Al_2O_3 + 6HCl \qquad (8.15')$$

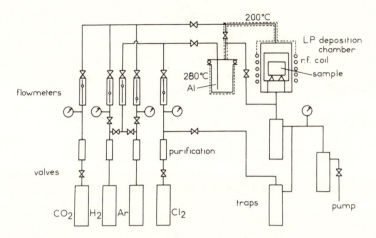

Figure 8.25 Experimental set-up for the infiltration of ceramic fibrous preforms by alumina deposited from a $AlCl_3$–H_2–CO_2 precursor (after [52]).

Detailed thermodynamic analyses [48, 49] have shown that the oxide is formed as a pure single phase on an inert substrate, provided water is in excess at equilibrium, that is when the P_{H_2}/P_{CO_2} ratio in the gas phase falls between two critical values α_i and α_s as shown in Figure 8.26 for zirconia. For hydrogen-rich compositions ($\alpha > \alpha_s$), the deposits no longer consist of pure oxides but of carbon/oxide mixtures, free carbon resulting from the reduction of carbon monoxide by hydrogen in excess, according to the following equation:

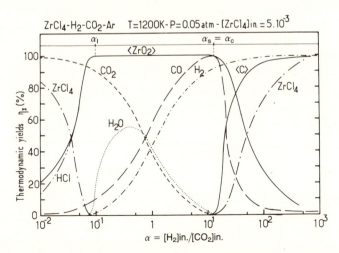

Figure 8.26 Influence of the precursor composition (α-ratio) on the thermodynamic yields in the CVD of zirconia on a chemically inert substrate (after [49]).

$$CO + H_2 \rightleftharpoons C + H_2O \qquad (8.16)$$

Finally, under conditions leading to pure oxide deposits (with a 100% yield), a carbon substrate (e.g. a carbon fibre preform) is etched by the vapour phase at the very beginning of the deposition process whereas a mullite substrate (e.g. a preform made of alumina-silica fibres) remains unattacked.

The deposition of alumina from $AlCl_3$–H_2–CO_2 and zirconia from $ZrCl_4$–H_2–CO_2 remain rate-controlled by surface reaction kinetics within the 800–1200 °C temperature range already considered for non-oxide ceramics. No evidence of transition to a regime where the process is rate-controlled by diffusion mass transfer has been yet reported [50, 51].

Various kinds of fibre preforms have been used to establish the feasibility of the ICVI process for both alumina and zirconia matrices including: (i) preforms made of Al_2O_3–SiO_2 short fibres (either as mats or randomly orientated) consolidated with a small amount of alumina or silico-alumina binder according to a wet slurry impregnation/firing procedure, (ii) 2D-preforms made of a stacking of Al_2O_3–SiO_2; Al_2O_3–SiO_2–B_2O_3 or carbon fabrics as well as pseudo-3D carbon fibre architectures, consolidated with a CVI binder of pyrocarbon or boron nitride and finally (iii) dry preforms consisting of FP corundum fibres maintained aligned unidirectionally with a ceramic tool. These preforms differ from one another in fibre volume fraction ($0.10 < V_f < 0.40$), overall initial porosity ($0.35 < V_{po} < 0.85$) and pore spectrum [52, 53]. Although the CVI parameters should have been optimised for each type of preform, as already discussed in section 8.2, the following ICVI conditions were used for a laboratory scale apparatus: $T = 900$–1000 °C; $P = 2$–3 kPa and gas flow rate $= 100 \text{ cm}^3 \text{ min}^{-1}$ for both alumina and zirconia ICVI [52, 54].

Examples of infiltration kinetic curves are shown in Figure 8.27 for the preforms and ICVI conditions presented above, for zirconia. The infiltration parameters were acceptable for all preforms at the beginning of the densification process. However, this was no longer true for some of them near the end of the process (as supported by deviation of data points from a linear $\ln(V_p/V_{po}) = f(t)$ relation). Residual porosities similar to those reported for non-oxide matrix CVI composites, that is of the order of 8–15%, were achieved for both alumina and zirconia [54].

The alumina and zirconia matrices deposited under the CVI conditions mentioned above, are often black materials due to small amounts of free carbon (and possibly to deviation from stoichiometry). The alumina CVI matrix consists mainly of corundum (α-Al_2O_3). Zirconia matrix films deposited under the same conditions but on flat non-porous substrates can contain percentages of the tetragonal modification as high

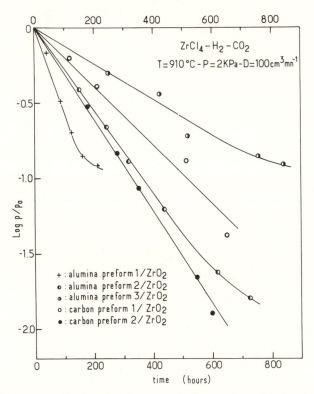

Figure 8.27 Infiltration kinetics of various ceramic fibre preforms by zirconia (deposited from a $ZrCl_4$–H_2–CO_2 precursor) (after [53]) *preform 1*: Saffil alumina fibre mat consolidated with an alumina binder ($V_{po} = 60\%$); *preform 2*: short alumina fibres (Zircar) bonded with a Al_2O_3–SiO_2 slurry after firing ($V_{po} = 85\%$); *preform 3*: 2D-alumina-silica preform (Sumitomo fibres) consolidated with CVI-BN ($V_{po} = 53\%$); *preform 4*: 2D-C (ex-PAN) preform consolidated with pyrocarbon ($V_{po} = 50\%$); and *preform 5*: 3D-C (ex-PAN) Novoltex preform from SEP consolidated with pyrocarbon ($V_{po} = 70\%$).

as 40%. This latter modification has not been observed in zirconia CVI matrices [51, 54].

8.4.2 Properties of oxide-matrix CVI composites

8.4.2.1 Mechanical behaviour at room temperature. The mechanical properties of both alumina and zirconia-matrix composites, derived from either alumina or carbon fibre preforms and loaded in compression or flexure at room temperature, have been the subject of preliminary studies [52, 53]. They strongly depend on: (i) the degree of densification of the preform (i.e. on the volume fraction of oxide and residual porosity), (ii) the applied load direction (particularly for the 2D-preform), (iii) the nature of the reinforcement and (iv) the fibre-matrix bonding (presence or not of a compliant interphase).

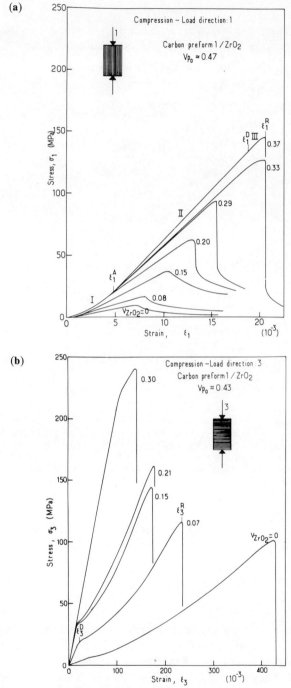

Figure 8.28 Compression behaviour at room temperature of 2D-C–C/ZrO₂ composites processed by CVI from 2D-C–C preforms, at various stages of pore filling: (a) load applied along direction 1, (b) load applied along direction 3 (after [53]). Nature of the preform: stacked 2D-carbon fabric (ex-PAN fibres) consolidated with pyrocarbon.

As shown in Figure 8.28, the mechanical behaviour *in compression* of 2D-C–C/ZrO_2 composites is similar to that already reported for non-oxide matrix composites derived from the same preforms. The non-linear stress–strain domain, related to the microcracking of the brittle zirconia matrix is very limited or very extended, depending on whether the load is applied along directions 1 or 3, respectively. Furthermore, the strain at failure is at least one order of magnitude higher, for a given degree of densification, when load is applied in direction 3. As seen in Figure 8.29, the stiffness and failure strength both increase with the volume fraction of matrix deposited in the pores of the preform, the material becoming stronger as the residual porosity becomes lower. Parabolic and exponential relations have been proposed to depict the variations of Young's moduli and failure stress as a function of V_{ZrO_2} or V_p [53].

As shown in Figures 8.30 and 8.31, the mechanical behaviour in flexure (3 point bending) is strongly dependent on the nature of the fibre-matrix bonding. Composites derived from preforms in which alumina based fibres are bonded together with an alumina or silicoalumina-binder (Figure 8.30), behave as brittle ceramics with: (i) stress–strain curves remaining linear almost up to failure and (ii) a low strain to failure, assumed to be related to the stress level at which the brittle matrix undergoes microcracking, that remains constant over a wide V_{ZrO_2} range when load is applied in direction 3. In contrast composites made from preforms in which the fibres are bonded together with a BN-binder (Figure 8.31) exhibit the typical features of tough ceramic composites, namely: (i) an extended domain of non-linear stress–strain relationship prior to failure related to damaging mechanisms, (ii) a load bearing capability which remains high after the first fibre failure (known to be due to fibre pull-out phenomena) and finally (iii) a high strain to failure. This difference in mechanical behaviour can be explained by the fact that, in the first case, the fibres are strongly bonded to the matrix possibly by a strong interphase (i.e. the alumina or silico-alumina binder) whereas, in the second case, they are bonded to the matrix with a *compliant interphase* (i.e. the BN binder) which acts as a mechanical fuse protecting the brittle fibres against the notch effect arising from the microcracking of the zirconia matrix [53].

To the author's knowledge no characterisation of oxide-matrix CVI composites loaded in tension has yet been reported.

8.4.2.2 Effect of environmental parameters. Although oxide-matrix composites have been designed for applications in air, all the data currently available on their *high-temperature mechanical properties* were obtained from tests performed under inert atmospheres (i.e. argon with 5 vol% hydrogen).

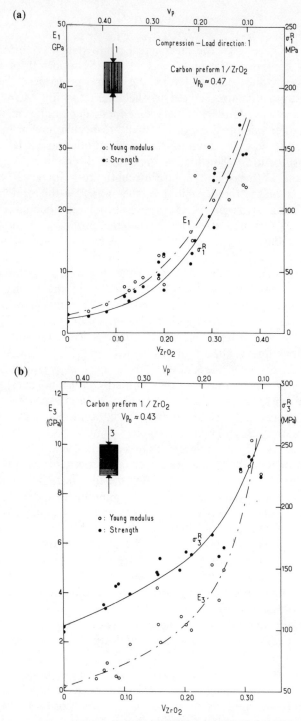

Figure 8.29 Variations of the Young's modulus E and the stress to failure σ^R of 2D-C–C/ZrO$_2$ composites as a function of the zirconia volume fraction, in compression loading: (a) load applied along direction 1, (b) load applied along direction 3 (after [53]).

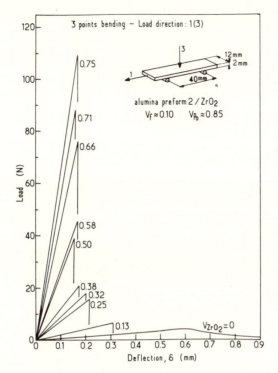

Figure 8.30 Bending behaviour at room temperature of ZrO_2-matrix composites processed by CVI from an Al_2O_3 based preform (alumina-silica fibres from Zircar, bonded with a silicoaluminous binder (after [53]).

Alumina (Saffil fibres)/zirconia composites ($V_f = 0.17$; $V_{ZrO_2} = 0.44$; $V_p \approx 0.15$; alumina binder) undergo a 5% decrease in stiffness (E_1) when heated to 1200 °C, for the first time. The composites then retain this stiffness when cooled to room temperature and reheated [53].

The variation of the flexural strength, as a function of temperature, exhibits the same general features for both alumina/alumina and alumina/zirconia composites (Figure 8.32). For composites of 17 vol% Al_2O_3–3% SiO_2 fibres (Saffil) bonded with an alumina binder, the failure strength is low and remains almost independent of temperature up to 1200–1300 °C. The failure strength is much higher for composites prepared from either corundum fibres (α-Al_2O_3 fibres; $V_f = 0.40$; dry preforms) or Al_2O_3–15% SiO_2 alumina fabrics (Sumitomo fibres; $V_f = 0.37$; BN binder), mainly due to the higher fibre volume fractions. However, for these two materials, a dramatic strength decrease occurs at about 1000 °C which must be due to an alteration of the fibre microstructure (i.e. crystallisation and/or grain growth). As a result, all the composites were reported to have almost the same residual failure strength at 1400 °C, that is about 50 MPa [52, 53]. The situation is only

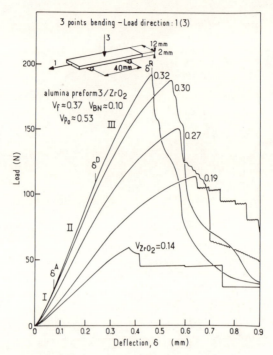

Figure 8.31 Bending behaviour at room temperature of ZrO_2-matrix composites processed by CVI from alumina based 2D-preforms (alumina-silica fabrics from Sumitomo bonded with BN deposited from a BF_3–NH_3 precursor) (after [53]).

slightly better for composites reinforced with carbon fibres. The residual strength at 1400 °C is higher (i.e. 100–150 MPa) due to the thermal stability and high strength of the carbon fibres but a strength decrease is still observed above about 1000 °C, for C–C/ZrO_2 composites, due presumably to chemical reactions taking place between the carbon preform and the zirconia matrix [53].

The *thermal expansion* and *thermal diffusivity* of alumina/alumina, alumina/zirconia and carbon/zirconia have been studied by Colmet *et al.* [52] and Minet *et al.* [55]. The thermal expansion of carbon/zirconia composites is of particular interest since: (i) the zirconia matrix, deposited as the monoclinic modification, is expected to undergo the monoclinic to tetragonal reversible phase transition with a specific volume change known for its detrimental effect in monolithic zirconia and (ii) the CTE of zirconia is positive and rather large whereas the axial CTE of carbon fibres is small and even negative. The thermal expansion of a pseudo 3D-C–C/ZrO_2 composite, in direction 1, is shown in Figure 8.33. The $ZrO_2^{(m)} \rightleftharpoons ZrO_2^{(t)}$ transition does occur in the CVI-zirconia matrix. However, it is more limited in amplitude with

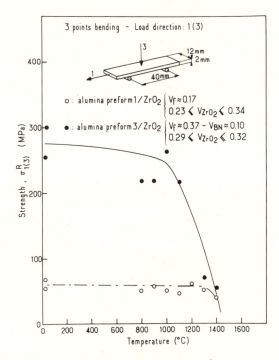

Figure 8.32 High-temperature bending strength of alumina/zirconia composites processed by CVI from a mat preform (Saffil fibres bonded with an alumina binder) and for a 2D-preform (Sumitomo fibres consolidated with BN deposited from a BF_3–NH_3 precursor) (after [53]).

respect to that for monolithic zirconia due to the presence of the carbon preform. Furthermore, since it takes place in zirconia cells of small volumes (i.e. the pores of the preforms), it has no significant effect on the integrity of the samples (the related microcracking of the matrix being an additional source of damaging). Finally, the overall thermal expansion, for example at 1000 °C is much lower than that reported for a monolithic zirconia (either unstabilised or partially stabilised) due to the contribution of the carbon fibres. Therefore, from a practical standpoint, the fibres in zirconia-matrix composites play a role similar to that of CaO, MgO or Y_2O_3 in stabilised monolithic zirconia, in overcoming the effects of the ZrO_2-phase transition.

The *thermal conductivity* of zirconia is known to be low. Furthermore, CVI composites have some residual porosity. As a result, ZrO_2-matrix composites are good thermal insulators. Alumina based fibres/zirconia CVI matrix composites ($V_p = 10$–15%) were reported to have a thermal conductivity similar (and even slightly lower) to that of dense sintered zirconia within the 1000–1500 °C temperature range. Similarly

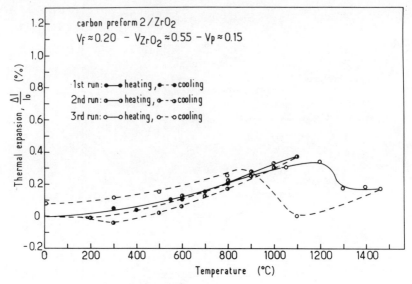

Figure 8.33 Thermal expansion of a 3D-C–C/ZrO_2 composite processed by CVI from a 3D-carbon preform (Novoltex fibrous architecture from SEP) consolidated with a pyrocarbon binder (deposited (from CH_4)) (after [55]).

pseudo-3D C–C/ZrO_2 composites (V_p = 8–10%) exhibit a thermal conductivity in the same temperature range as that of sintered alumina [55].

The *oxidation resistance* of oxide-matrix composites must be intrinsically better than that of non-oxide matrix composites. However, even for alumina/alumina or alumina/zirconia composites, a long exposure at high temperature in atmospheres containing oxygen may have several detrimental effects, for example on the mechanical strength: (i) a strength loss of the fibres due to the effect of temperature on their microstructure, (ii) an alteration of the microstructure of the matrix (e.g. oxidation of the small amount of carbon that it contains and grain growth) and (iii) an oxidation of the interphase material. Such effects have been observed for alumina fibres (Saffil)/zirconia composites, with a strength loss of 30–50%, after a treatment of 6 h at 1300 °C [55]. Obviously, the oxidation resistance must be weaker when the preform consists of carbon fibres. However, as shown in Figure 8.34, 2D-C–C/ZrO_2 composites undergo only a limited weight loss (i.e. of the order of 2%) when heated at 1100 °C in air, provided the volume fraction of infiltrated ZrO_2 is high enough (i.e. $V_{ZrO_2} > 0.30$) even if the residual porosity is rather high (of the order of 25%) [55]. This result clearly shows that a zirconia matrix, even if it is not a perfect diffusion barrier for oxygen, does protect a carbon preform against oxidation at

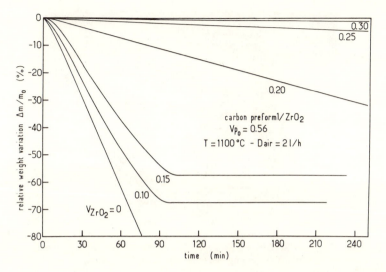

Figure 8.34 Kinetics of oxidation in air of 2D-C–C/ZrO₂ composites (processed by CVI from a 2D-C–C preform) at various stages of densification [55].

least for short exposures (e.g. a few hours) at moderate temperatures (about 1000 °C).

8.5 Applications

The main applications of *SiC-matrix composites* are in the field of the aerospace industry. C–SiC and/or SiC–SiC composites, made from 2D or pseudo-3D preforms by ICVI, are being evaluated for the following applications: (i) components for liquid propulsion rocket engines as well as for jet engines and (ii) reusable large dimension structures for spacecrafts (e.g. the European Hermes space shuttle). The advantages of SiC-matrix composites are a high oxidation resistance, with respect to carbon–carbon composites, and higher temperature capability as well as lower density when compared to refractory alloys. Examples of parts made of SiC-matrix composites, according to the ICVI process are shown in Figure 8.35 [20]. The use of SiC-matrix composites in prototypes of tubes, baskets, panels and heat exchangers has also been reported [56]. Finally, thick wall prototype tubes, made from wound tubular preforms and utilising the SiC-FCVI process, have been mentioned [7, 21]. However as far as the author is aware this technology has not been up-scaled to production level.

All the other refractory matrix composites which have been presented here remain experimental materials.

Figure 8.35   Examples of parts made of SiC CVI matrix composites designed to be used in advanced jet engines (courtesy of SEP-Bordeaux).

References

1. Stinton, D. P., Besmann, T. M. and Lowden, R. A. (1988) *Ceram. Bull.* 67, 350–355.
2. Naslain, R. and Langlais, F. (1986) in: *Tailoring Multiphase and Composite Ceramics*, eds. Tressler R. E. *et al., Mater. Sci. Res.* 20, 145–164.
3. Caputo, A. J., Lowden, R. A. and Stinton, D. P. (1985) ORNL/TM-9651, NTIS, US Dept. Commerce, Springfield, Virginia 22161, USA.
4. Broquère, B., Buttazzoni, B. and Choury, J. J. (1985) in: *Introduction aux Matériaux Composites, Vol. 2: Matrices Métalliques et Céramiques*, ed. R. Naslain, Coedition CNRS-IMC, Bordeaux, chap. 17, pp. 405–438.
5. Greniè, Y. (1987) in: *Looking Ahead for Materials and Processes*, eds. de Bossu J. *et al.*, Elsevier, Amsterdam, pp. 377–386.
6. Fitzer, E. and Hegen, D. (1979) *Angew. Chem. Int. Ed. Engl.* 18, 295–304.
7. Moeller, H. H., Long, W. G., Caputo, A. J. and Lowden, R. A. (1987) *Ceram. Eng. Sci. Proc.* 8, 977–984.
8. Dugne, O., Prouhet, S., Guette, A., Naslain, R. and Sevely, J. (1989) *Proc. ECCM-3 Developments in the Science and Technology of Composite Materials* eds. Bunsell A. R. *et al.*, Bordeaux, Elsevier Applied Science, London, pp. 129–135.
9. Lowden, R. A. and Stinton, D. P. (1987) ORNL/TM-10667, NTIS- US Dept. Commerce, Springfield, Virginia 22161, USA
10. Fitzer, E. and Gadow, R. (1986) *Ceram. Bull.* 65, 326–335.
11. Fitzer, E., Hegen, D. and Strohmeier, H. (1980) *Rev. Int. Hautes Temper. Refract.* 17, 23–32.
12. Naslain, R., Rossignol, J. Y., Hagenmuller, P., Christin, F., Heraud, L. and Choury, J. J. (1981) *Rev. Chim. Minérale* 18, 544–564.
13. Rossignol, J. Y., Langlais, F. and Naslain, R. (1984) *Proc. 9th Int. Conf. CVD*, eds. Robinson, Mc.D. *et al.*, Cincinnati, The Electrochem. Soc., Pennington, NJ, pp. 596–614.
14. Naslain, R., Langlais, F. and Fedou, R. (1989) *J. Physique* 50, C5.191–C5.207.
15. Fitzer, E., Fritz, W. and Gadow, R. (1983) *Proc. Adv. Ceram. Mater.*, Tokyo Institute of Technology, Yokohama.
16. Gupte, S. M. and Tsamopoulos, J. A. (1989) *J. Electrochem. Soc.* 136, 555–561.

17. Van den Brekel, C. H. J., Fonville, R. M. M., Van der Straten, P. J. M. and Verspui, G. (1981) *Proc. 8th Int. Conf. CVD*, Paris, eds. Blocher, J. M. *et al.*, The Electrochem. Soc., Pennington, NJ, pp. 142–156.
18. Christin, F., Naslain, R. and Bernard, C. (1979) *Proc. 7th Int. Conf. CVD*, Los Angeles, eds. Sedwick, T. O. and Lydin, H., The Electrochem. Soc., Princeton, NJ, pp. 499–514.
19. Christin, F. (1979) Thesis # 641, University of Bordeaux.
20. Cavalier, J. C., Lacombe, A. and Rouges, J. M. (1989) *Proc. ECCM-3*, Bordeaux, eds. Bunsell, A. R. *et al.*, Elsevier Applied Science, London, pp. 99–110.
21. Caputo, A. J., Lackey, W. J. and Stinton, D. P. (1985) *Ceram. Eng. Sci. Proc.* 6, 694–706.
22. Stinton, D. P., Caputo, A. J. and Lowden, R. A. (1986) *Ceram. Bull.* 65, 347–350.
23. Starr, T. L. (1987) *Proc. 10th Int. Conf. CVD*, Honolulu, eds. Cullen, G. W. The Electrochem. Soc., Pennington, NJ, pp. 1147–1155.
24. Tai, N. H. and Chou, T. W. (1989) *Am. Ceram. Soc. Annual Meeting*, Indianapolis.
25. Moeller, H. H., Long, W. G., Caputo, A. J. and Lowden, R. A. (1986) *SAMPE Quarterly* 17, 1–4.
26. Bernhart, G., Lamicq, P. and Mace, J. (1985) *L'industrie Céramique* 790, 51–56.
27. Gomina, M., Chermant, J. L. and Osterstock, F. (1985) in: *Adv. Mater. Res. and Development for Transportation Composites*, eds. Lamicq, P. *et al.*, MRS-Europe, Strasbourg, Les Editions de Physique, Les Ulis, France, pp. 117–124.
28. Gomina, M., Chermant, J. L., Osterstock, F., Bernhart, G. and Mace, J. (1986) in: *Fracture Mechanics of Ceramics*, Vol. 7, eds. Bradt, R. C., Hasselman, D. P. H., Evans, A. G. and Lange, F. F., Plenum Press, New York and London, pp. 17–32.
29. Bouquet, M., Birbis, J. M. and Quenisset, J. M. (1990) *Composites Sci. Technol.* special issues on Ceramic Matrix Composites 37, 223–248.
30. Abbe, F., Chermant, L., Coster, M., Gomina, M. and Chermant, J. L. (1990) *Composites Sci. Technol.*, special issues on Ceramic Matrix Composites 37, 109–127.
31. Lamicq, P. J., Bernhart, G. A., Dauchier, M. M. and Mace, J. G. (1986) *Ceram. Bull.* 65, 336–338.
32. Dauchier, M. M., Bernhart, G. A. and Bonnet, C. (1985) *Proc. 30th Nat. SAMPE Symp.*, Covina, CA, pp. 1519–1525.
33. Bouquet, M., Birbis, J. M., Quenisset, J. M. and Naslain, R. (1987) *Proc. ICCM-6/ECCM-2* eds., Matthews, F. L. *et al.*, Elsevier Applied Science, London, Vol. 2, pp. 48–59.
34. Gomina, M., Chermant, J. L. and Osterstock, F. (1984) *Proc. 2nd Int. Conf. Creep Fracture of Engineering Mater. and Structures*, eds. Wilshire, B. and Owen, D. R. J., Pinebridge Press, Vol. 1, pp. 541–550.
35. Lancin, M., Anxionnaz, F., Schuhmacher, M., Dugne, O. and Trebbia, P. (1987) *Mat. Res. Soc. Symp. Proc.*, The Materials Research Society, Vol. 78, pp. 231–238.
36. Naslain, R., Hagenmuller, P., Christin, F., Heraud, L. and Choury, J. J. (1980) *Proc. ICCM-3 Advances in Composite Materials*, Paris, eds. Bunsell, A. R. *et al.*, Pergamon Press, Oxford, Vol. 2. pp. 1084–1097.
37. Prewo, K. M. and Batt, J. A. (1988) *J. Mater. Sci.* 23, 523–527.
38. Grande, D. H., Mandell, J. F. and Hong, K. C. C. (1988) *J. Mater. Sci.* 23, 311–328.
39. Hannache, H., Langlais, F. and Naslain, R. (1985) *Proc. 5th European Conf. CVD, EURO-CVD-5*, eds. Carlsson, J. O. and Lindström, J., Uppsala, Uppsala Univ. Press, pp. 219–233.
40. Pierson, H. O. (1975) *J. Composite Mater.* 9, 228–240.
41. Hannache, H., Naslain, R. and Bernard, C. (1983) *J. Less-Common Metals* 95, 221–246.
42. Hannache, H., Quenisset, J. M., Naslain, R. and Heraud, L. (1984) *J. Mater. Sci.* 19, 202–212.
43. Fitzer, E. (1978) *Proc. Int. Symp. Factors Densification and Sintering of Oxide and Non-oxide Ceram.*, Hakone, Japan, pp. 40–76.
44. Rossignol, J. Y., Quenisset, J. M., Hannache, H., Mallet, C., Naslain, R. and Christin, F. (1987) *J. Mater. Sci.* 22, 3240–3252.

45. Rossignol, J. Y., Naslain, R., Hagenmuller, P. and Heraud, L. (1982) *Proc. ICCM-4 Progress in Science and Engineering of Composites*, eds. Hayashi, T. *et al.*, Tokyo, pp. 1227–1237.
46. Rossignol, J. Y. (1985) Thesis # 833, Univ. Bordeaux, France.
47. Hannache, H. (1984) Thesis # 813, Univ. Bordeaux, France.
48. Lhermitte-Sebire, I., Colmet, R., Naslain, R. and Bernard, C. (1986) *J. Less-Common Metals* 118, 83–102.
49. Minet, J., Langlais, F., Naslain, R. and Bernard, C. (1986) *J. Less-Common Metals* 119, 219–235.
50. Colmet, R., Naslain, R., Hagenmuller, P. and Bernard, C. (1982) *J. Electrochem. Soc.* 129, 1367–1372.
51. Minet, J., Langlais, F. and Naslain, R. (1987) *J. Less-Common Metals* 132, 273–287.
52. Colmet, R., Lhermitte-Sebire, I. and Naslain, R., (1986) *Adv. Ceram. Mater.* 1, 185–191.
53. Minet, J., Langlais, F., Quenisset, J. M. and Naslain, R. (1989) *J. Europ. Ceram. Soc.* 5, 341–356.
54. Minet, J., Langlais, F. and Naslain, R. (1991) *Composites Sci. Technol.* 37, 79–107.
55. Minet, J., Langlais, F. and Naslain, R. (1991) *J. Europ. Ceram. Soc.* 7, 283–294.
56. Reagan, P., Ross, M. F. and Huffman, F. N. *Adv. Ceram. Mater.* 3, 198–201.

9 Whisker reinforced ceramics

J. HOMENY

9.1 Introduction

High performance polycrystalline ceramics, such as Si_3N_4, SiC, Al_2O_3, and ZrO_2, exhibit a combination of good erosion and corrosion resistance, high hardness, and high elastic modulus at elevated temperatures. However, these structural ceramics are brittle, with low fracture toughnesses and probabilistic fracture stresses determined by their inherent flaw population. While design methodology has improved substantially over the years, the tendency for these brittle ceramics to fail in a catastrophic manner has remained the major factor limiting their use in structural applications.

In recent years, there has been considerable interest in utilising ceramic whiskers for reinforcing polycrystalline ceramics, as the whisker reinforcement mitigates the problem of brittleness and unpredictable fracture stress. The success of these ceramic-matrix composites depends principally on obtaining high mechanical reliability, the combination of improved fracture toughness and fracture stress. The rationale for the current interest is that these composites offer the greatest potential as structural ceramics, due to whisker pull-out and bridging toughening mechanisms. Other toughening mechanisms, such as microcracking, crack deflection and phase transformation, all suffer serious limitations. To date, whisker reinforced ceramics have been fabricated with fracture toughnesses in the $5-15\,MPa\,m^{1/2}$ range, where structural applications become feasible.

9.2 Composite systems

The main goal in the fabrication of ceramic-matrix composites is to significantly increase the fracture toughness, while maintaining the other attractive high temperature properties, for example fracture stress [1]. In order that a particular whisker/matrix combination will form a successful composite, three factors must be considered; (1) whisker/matrix thermal expansion mismatches, (2) whisker/matrix elastic moduli mismatches, and (3) whisker/matrix chemical compatibility during the high temperature densification process. The ability to design composites with desired properties is also directly related to a control of critical microstructural features. To a large extent, the type, size, shape, volume fraction, distribution and surface characteristics of the whiskers dictate the development of the microstructure. Thus, the incorporation of whiskers into polycrystalline ceramics can modify the development of microstructural features, especially when significant chemical interaction occurs between the whiskers and matrix. Therefore, a fundamental understanding of the relationship between microstructure and fabrication parameters is essential for successful development of composites with enhanced properties.

9.2.1 *Whiskers*

Recent advances in ceramic-matrix composites have been made possible, in part, by the availability of ceramic whiskers in large quantities at a relatively moderate cost. The whiskers, primarily SiC, are commonly made by either the carbothermal reduction process, that is the rice hull process, or by the recently developed vapor-liquid-solid process. In the former, rice hulls are calcined at approximately 900 °C, prior to pyrolysis at 1700 °C to form SiC whiskers [2]. For this process, the production cost is modest and there are no intrinsic difficulties in scaling production to large quantities. Typical whisker morphology for rice hull derived SiC whiskers, is illustrated in Figure 9.1. In the vapor-liquid-solid process, hydrocarbons (e.g. CH_4) and SiO feed vapours are reacted to approximately 1400 °C, via a molten metal catalyst, to precipitate SiC whiskers [3]. For this process, although production quantities are currently small

Figure 9.1 Scanning electron microscope micrograph of rice hull derived SiC whiskers (Silar SC-9 ARCO Metals Company) illustrating typical length and diameter variability (bar = 10 μm) (Vaughn *et al.* [47]).

and prices are quite high, the prognosis for modest cost whiskers is also good.

Of great importance is the intrinsic thermal and chemical stability of the ceramic whiskers. They are single crystals of stoichiometric compounds, with no tendency for recrystallisation, internal chemical reactions, or other detrimental processes. The SiC whiskers produced from the rice hull process are typically less than 1 μm in diameter and range from 10 to 50 μm in length. They are characterised by a structure that is heavily faulted, consisting of planar defects in the close-packed planes perpendicular to the growth direction; that is the long axis of the whisker [4, 5, 6]. The planar defects are a result of the arrangement of regions of the α and β polytypes of SiC. Impurities, for example Al, Mn, Mg, Fe and Ca, and internal voids can also be present in the SiC whiskers. Furthermore, the surface chemistry tends to differ from that of the bulk, in that the surfaces tend to be oxygen rich [4, 7]. Surface species commonly identified resemble either crystalline SiO_2 or amorphous Si–O–C [8]. The vapor-liquid-solid process is capable of producing whiskers in a variety of microstructures, but the typical dimensions are 5–6 μm in diameter and up to 100 mm in length. They are usually in the β form of SiC and exhibit a lesser degree of internal defects than rice hull whiskers [9].

SiC whiskers are the most promising reinforcements to date, as they retain their mechanical properties up to at least 1600 °C. Selected properties of SiC whiskers, along with other types of ceramic whiskers, are illustrated in Table 9.1. Fracture stresses are not included since these measurements are difficult to perform. However, the fracture stress of 5 mm long vapor-liquid-solid SiC whiskers was found to be approximately 8 GPa [9]. Since most whiskers are shorter than 1 mm, this value is a reasonable estimate for the fracture stress.

9.2.2 *Matrices*

It is possible to prepare whisker reinforced ceramics with good thermal and chemical stability, provided that the whiskers are combined with the appropriate ceramic matrix, chosen in order to minimise interdiffusion and interfacial reactions. Because of the high surface areas and small diffusion distances involved with the use of whisker reinforcements, chemical compatibility is essential in maintaining composite properties. For SiC whiskers, Al_2O_3, Si_3N_4, mullite ($3Al_2O_3 \cdot 2SiO_2$), and ZrO_2 have been primarily utilised as matrices [10]. Selected properties of these ceramic matrices, along with other types of matrices, are illustrated in Table 9.2. Except for the glass matrices, the microstructures are polycrystalline with average grain sizes ranging from approximately

Table 9.1 Typical properties of ceramic whiskers based on data of manufacturers

Whisker	Diameter (μm)	Length (μm)	Young's elastic modulus (GPa)	Thermal expansion coefficient (cm/cm °C)	Density (g/cm³)
SiC[a]	3–10	500–10 000	580	4.5×10^{-6}	3.2
SiC[b]	0.5–0.6	10–80	600	4.5×10^{-6}	3.2
SiC[c]	0.1–1.0	50–200	600	4.5×10^{-6}	3.2
SiC[d]	0.05–0.2	10–40	600	4.5×10^{-6}	3.2
SiC[e]	0.05–1.0	5–100	600	4.5×10^{-6}	3.2
Si_3N_4[f]	0.05–0.5	5–100	370	2.5×10^{-6}	3.2
Al_2O_3[g]	4–7	40–100	400	7×10^{-6}	4.0

[a] Vapor-liquid-solid, Los Alamos National Laboratory.
[b] Silar SC-9, ARCO Metals Company.
[c] Tokamax, Tokai Carbon Company.
[d] Tateho SCW-1, Tateho Chemical Industries Company.
[e] Huber XPW2, Huber Company.
[f] UBE-SNWB, UBE Industries.
[g] Catapal XW, Vista Chemical Company.

Table 9.2 Typical properties of ceramic matrices

Ceramic matrix	Young's elastic modulus (GPa)	Thermal expansion coefficient (cm/cm °C)	Fracture stress (MPa)	Fracture toughness (MPa m$^{1/2}$)	Density (g/cm^3)	References
Al_2O_3	400	8.9×10^{-6}	350–700	3–5	3.97	[11]
Si_3N_4	420	2.9×10^{-6}	400–900	4–6	3.26	[11, 12]
ZrO_2[a]	240	13.5×10^{-6}	150–350	2.5	5.6–6.1	[13]
$3Al_2O_3 \cdot 2SiO_2$	145	5.3×10^{-6}	180–200	2–3	3.16	[11]
Glass	60–80	$3–10 \times 10^{-6}$	70–100	1–2	2.3–3.0	[13]
Glass ceramics	80–120	$1.5–17 \times 10^{-6}$	50–350	2–3	2.4–5.9	[13]

[a]Fully stabilised

1–50 μm. Both the matrix grain size and morphology affect the fracture mode, that is intragranular versus intergranular, and thus have significant effects on the mechanical properties. Therefore, matrix microstructure must be factored into the composite design.

9.3 Composite processing

Before discussing individual stages in composite preparation processes, a brief outline of general principles is presented.

Two important objectives during composite processing are (1) attainment of uniform distribution of whiskers and (2) minimisation of mechanical and/or chemical damage to whiskers. These objectives must be considered during both the preform fabrication, that is powder processing, and the densification processes. Preform fabrication is commonly performed at room temperature, while densification occurs at high temperatures, for example 1000–2000 °C.

Processing methods for whisker reinforced ceramics have developed from established methods used to process monolithic ceramics. An important advantage of whisker reinforced ceramics over continuous fibre reinforcement is that these well known processing techniques can be applied. However, there are processing problems associated with the physical character of the whiskers, for example small diameter, high surface area, high aspect ratio and high reactivity. Since whiskers are harvested and handled as dry ceramic powders, a high degree of whisker agglomeration is common. This leads to difficulties in dispersing whiskers and mixing them uniformly with matrix powders. Typical processing flaws observed in composites are primarily clusters of whiskers devoid of matrix and matrix regions devoid of whiskers. Additional residual contamination in whiskers, for example particulates, can also provide potential failure origins in composites. However, these potential flaw types can be eliminated by flotation and sedimentation techniques.

Blends of whiskers and matrix powders can usually be processed by conventional means. A typical processing procedure consists of the following: (1) whisker purification by sedimentation to remove whisker agglomerates and particulates, (2) mixing of matrix powders and appropriate sintering aids by milling, (3) mixing of matrix powders and whiskers by high speed blending and/or ultrasonic dispersion in a liquid medium, (4) drying to remove the liquid medium, (5) shape forming by cold pressing, injection molding, or slip casting, and (6) densification by hot pressing or sintering and/or hot isostatic pressing. If preform fabrication is achieved by slip casting then the fourth stage of liquid removal would not be required. Hot pressing yields a composite with low porosity, but is not suitable for the fabrication of complex shapes.

Achieving low porosity composites by sintering is difficult because the whiskers inhibit densification. If composites of complex shapes can be sintered to the state where only closed porosity exists, then final densification can be achieved by hot isostatic pressing.

The fabrication technique employed is strongly related to the degree of whisker orientation. Various techniques have been proposed for achieving whisker orientation. For one dimensional uniaxially aligned whiskers (i.e. orthotropic composites), the primary obstacle is achieving alignment in the whisker preform. The one dimensional composites provide the least constraint to consolidation and are the most amenable to densification via pressureless sintering. For two dimensional random whisker orientation (i.e. transversely isotropic composites), conventional processing techniques can be used, for example mixing plus hot pressing. Whisker preforms may also be produced by slip casting or centrifugal casting from a whisker suspension incorporating binders; such preforms can then be infiltrated with matrix precursors. For three dimensional random whisker orientation, that is macroscopically isotropic composites, there are currently no suitable fabrication techniques available, either for producing the three dimensional whisker orientation or for consolidating the composites.

9.3.1   *Whisker purification*

Whiskers contain varying amounts of particulate impurities, usually in the form of large crystals and bundles of whiskers [14]. Large particulates are particularly detrimental because they act as strength limiting flaws. Sedimentation techniques are effective in removing these particulates. A sedimentation-flocculation technique, developed by Lundberg *et al.* [15], removes the particulates without recreating agglomerates in the drying step. The zeta potential of the whiskers is measured to determine what pH is needed for dispersion or flocculation. Whiskers are then dispersed at a high zeta potential, that is positive or negative. Stoke's equation is applied to calculate sedimentation times for particles of a certain size:

$$v = [2r^2(\rho_s - \rho_l)g/9\eta]$$

where v = sedimentation velocity, ρ_s = density of the solid, ρ_l = density of the liquid, η = liquid viscosity, and r = equivalent spherical radius. Sedimentation is carried out at the calculated times and the supernatant liquid is removed in order to obtain the desired size fractions. The individual size fractions are then flocculated at a zeta potential near zero. A loose sediment is formed, as the flocculated whiskers settle rapidly. The sediment can be partially dewatered by centrifuging and stored in the wet state to prevent hard agglomerates from forming. In this manner, both the fine and coarse particulate material can be eliminated.

9.3.2 Whisker/matrix powder mixing

The use of wet processing techniques for whisker/matrix powder mixing is most desirable because it ensures a homogeneous distribution of the whiskers. Depending on the method to be used to fabricate the composite preform, wet mixing can be carried out in either a dispersed or a flocculated state. For composite preform fabrication by slip casting, a well stabilised and dispersed slip with low viscosity and high solids content is necessary to achieve high preform densities. A well dispersed slip is obtained by choosing a pH where both the matrix powders and whiskers have a high zeta potential of the same charge. Surfactants can also be employed to increase the slip stability. Alternate fabrication approaches include pressure or centrifugal slip casting in order to produce compacts of higher density. For composite preform fabrication by cold pressing, a flocculated state is necessary. Flocculation occurs when the whiskers and matrix powders are either differently charged or have zeta potentials near zero. To effectively eliminate agglomerates and whisker clusters the flocculated system must be subjected to high shear mixing in order to achieve a homogeneous mixture. Examples of whisker/matrix powder mixing procedures are illustrated in Figure 9.2a–c.

Sintering aids are typically added during the whisker/matrix powder mixing process. The sintering aids promote densification by enhancing solid state diffusion and/or liquid phase sintering processes. For Si_3N_4 matrices, Y_2O_3, CeO, Al_2O_3, and MgO have been found to be effective sintering aids [4, 12, 18] while for Al_2O_3 matrices, MgO and Y_2O_3 have been used successfully [19]. When adding these sintering aids, their chemical compatibility with the whiskers must also be considered.

9.3.3 Densification

After composite preform fabrication, densification is achieved by hot pressing, pressureless sintering, hot isostatic pressing or a combination

---

Figure 9.2 (a) Whisker/matrix powder processing in preparation for composite preform fabrication by slip casting (Hoffmann *et al.* [16]). (b) Whisker/matrix powder processing in preparation for composite preform fabrication by cold pressing (Homeny *et al.* [17]). (c) Whisker/matrix powder processing in preparation for composite preform fabrication by cold pressing (Buljan *et al.* [18]).

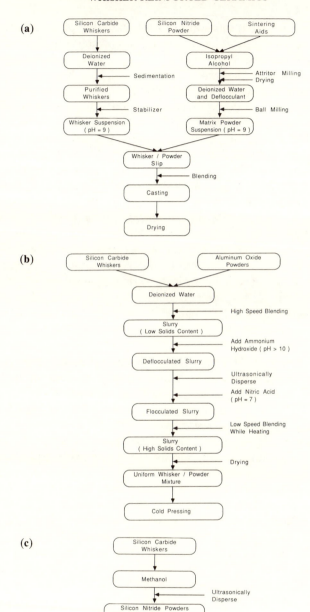

of pressureless sintering followed by hot isostatic pressing. The hot pressing technique is capable of producing composites with near-theoretical density. Pressureless sintering may also produce useful composites, but substantially below theoretical density. Uniaxial hot pressing tends to produce significant whisker orientation normal to the pressing direction, with whiskers randomly distributed on planes normal to this direction, that is transversely isotropic composites. The extent of orientation is clearly a function of die geometry, whisker aspect ratio and matrix characteristics. Becher and Wei [20] have documented the magnitude of the fracture stress and fracture toughness anisotropy in SiC whisker/Al_2O_3 matrix composites as a function of whisker orientation. Consolidation via sintering is limited by the constraints of the whiskers, which are rigid and do not shrink during sintering. Whiskers are also effective grain growth inhibitors. Homeny et al. [17] and Becher et al. [21] have observed decreasing matrix grain size with increasing whisker contents in SiC whisker/Al_2O_3 matrix composites and Buljan et al. [18] have observed similar grain growth inhibition in SiC whisker-/Si_3N_4 matrix composites. Hot isostatic pressing is usually used in conjunction with pressureless sintering, provided a state of closed porosity can be achieved during the sintering process.

9.3.3.1 Hot pressing. Hot pressing is currently the only viable process for producing dense composites. Whisker/matrix powder mixtures are typically pressed at temperatures from 1000 to 2000 °C in carbon or graphite dies. Hot pressing times and pressures are of the order of 30–300 min and 30–70 MPa, respectively. The use of a vacuum or inert atmospheres (e.g. argon), during hot pressing are necessary to prevent oxidation of the carbon or graphite dies and non-oxide components of the composites.

Numerous researchers have examined the effect of hot pressing parameters on the densification of whisker reinforced ceramics. Wei and Becher [11] performed extensive densification studies on SiC whisker/Al_2O_3 matrix composites. They found the densification rate was a function of hot pressing temperature and pressure. Homeny et al. [17] examined the effect of hot pressing temperature and time on the densification of the same composite system. They found that the percent theoretical density increased with increasing time at a constant temperature or with increasing temperature at a constant time. However, it was noted that at the highest temperatures employed slight decreases in density were observed, possibly due to loss of material through the vapour phase. Buljan et al. [18] examined the densification rate as a function of whisker content for SiC whisker/Si_3N_4 matrix composites. They found that the densification rate decreased with increasing whisker content and concluded that the whiskers substantially inhibited grain rearrangement.

9.3.3.2 Pressureless sintering/hot isostatic pressing. Fabrication of dense composites with high whisker contents by pressureless sintering alone is difficult; some form of encapsulation and pressure assistance appears to be necessary. Densification of cold pressed or slip cast compacts is inhibited by the rigid whiskers because grain rearrangement is difficult during the densification process. Essentially, the whiskers form a network structure, which severely inhibits matrix shrinkage. The matrix powder tends to sinter into small aggregates located at the intersection of whiskers, effectively locking the structure at the whisker intersection nodes, therefore preventing significant densification. Moreover, when whiskers are introduced into a matrix powder, efficient particle packing is prevented, resulting in low preform densities. Thus, to promote sintering, control of whisker aspect ratio, matrix particle size, preform density and sintering aids is necessary. Typical sintering and hot isostatic pressing temperatures range from 1000 to 2000 °C. Pressures up to 200 MPa are employed during hot isostatic pressing.

Several investigators have fabricated composites by sintering and/or hot isostatic pressing techniques. Tiegs and Becher [19] have demonstrated that SiC whisker/Al_2O_3 matrix composites with less than 10 vol% whiskers can be sintered to greater than 95% of theoretical density and subsequently hot isostatically pressed without encapsulation to greater than 98% of theoretical density. For whisker loadings greater than 10 vol%, densification was severely inhibited, since composites with open porosity were obtained. SiC whisker/Si_3N_4 matrix composites were slip cast and sintered by Hoffmann *et al.* [16]. They found that with increasing whisker content, from 0 to 20 vol%, sintered densities decreased from 98 to 88% of theoretical density. Pezzotti *et al.* [22] and Björk and Hermansson [23] used hot isostatic pressing successfully to densify SiC whisker/Si_3N_4 matrix and SiC whisker/Al_2O_3 matrix composites, respectively. Composite preforms were presintered at a relatively low temperature to remove organic binders used during cold pressing and encapsulated in glass prior to hot isostatic pressing.

9.4 Mechanical properties

Whisker reinforcements are primarily utilised to prevent catastrophic brittle failure by providing processes that dissipate energy during crack propagation. Numerous investigators have demonstrated that the fracture stress and/or fracture toughness of polycrystalline ceramics can be significantly improved by reinforcing them with single crystal whiskers. However, when whiskers are employed in two or three dimensional randomly oriented composites, the toughening and strengthening increments possible are limited to approximately 1/2 to 1/3 of that possible in aligned composites.

9.4.1 *Strengthening and toughening mechanisms*

Numerous investigators [21, 24–28] have reviewed the strengthening and toughening mechanisms relevant to fibre and whisker reinforced ceramics and some general principles are outlined in chapter 4. The important strengthening mechanisms relevant to whisker reinforcement are:

(1) *Load transfer*: significant load transfer strengthening requires $E_{\text{whisker}}/E_{\text{matrix}}$ ratios greater than 2, where $E_{\text{whisker}} = $ Young's elastic modulus of the whisker, and $E_{\text{matrix}} = $ Young's elastic modulus of the matrix. This mechanism also requires a strong whisker/matrix interfacial bond to transfer the load from the matrix to the high strength whiskers.

(2) *Matrix prestressing*: compressive stresses can develop in the matrix if $\alpha_{\text{whisker}} > \alpha_{\text{matrix}}$, where $\alpha_{\text{whisker}} = $ thermal expansion coefficient of the whisker and $\alpha_{\text{matrix}} = $ thermal expansion coefficient of the matrix. However, the development of large stresses can lead to interfacial debonding and microcracking, which lowers the fracture stress.

A number of different toughening mechanisms are probably operative to some extent in whisker reinforced ceramics, but some particular mechanism will dominate depending on the characteristics of the individual composite. Mechanisms which may have the potential to enhance the fracture toughness of whisker reinforced ceramics include:

(1) *Crack pinning*: crack pinning does not contribute significantly to the fracture toughness. This mechanism requires a small whisker size, high whisker volume fraction, strong whisker/matrix interfacial bond, and a whisker fracture energy greater than that of the matrix. These requirements are difficult to achieve in whisker reinforced ceramics because composites with whisker volume fractions above 30% are difficult to consolidate and the fracture energy of currently available whiskers is similar to that of the potential matrix materials.

(2) *Crack deflection*: crack deflection can contribute significantly to the fracture toughness, but calculations of toughening increments by this mechanism indicate a maximum potential increase of 50–100%. This mechanism forces the crack out of plane rather than through the whisker, that is the crack plane is no longer normal to the applied stress. Deflection results from residual stresses generated from either thermal expansion mismatches between the matrix and whiskers or from thermal expansion anisotropy in the whiskers, elastic mismatches between the matrix and whiskers which perturb local stress fields, or weak whisker/matrix interfacial bonding which offers the crack a low resistance

path. This mechanism also requires a high whisker volume fraction. Crack deflection has the beneficial effects of increasing fracture surface area, increasing the energy absorbed during fracture and forcing mixed mode fracture. The shift from mode I to modes II and III greatly increases the difficulty in crack propagation, since K_{IIc} and K_{IIIc} are typically much higher than K_{Ic} in brittle materials.

(3) *Microcracking*: microcracking can contribute significantly to the fracture toughness. This mechanism requires microcracks, either pre-existing or created during fracture, that decrease the stress intensity at the main crack tip. Microcrack generation requires either dispersed second phases with large thermal expansion anisotropy or large expansion mismatches between the dispersed second phases and the matrix. Large elastic mismatches between the dispersed second phases and the matrix may also interact with the stress field of the approaching crack and generate large stresses in the vicinity of the dispersed second phases. The main drawback to microcrack toughening is that it is limited to a narrow range of materials exhibiting significant anisotropy or mismatch.

(4) *Crack bridging*: crack bridging is a highly significant toughening mechanism. It requires that unfractured whiskers bridge the crack behind the crack tip. The closure forces exerted by the bridging whiskers reduce the crack opening displacement and the stress intensity at the crack tip. The operation of this mechanism generally requires strong whisker/matrix interfacial bonding, with a substantially higher strain to failure of the whiskers compared to the matrix. Whisker orientation is critical, as whiskers at a substantial angle to the normal to the crack plane will be subjected to shear and bending stresses and may not provide much bridging. Synergistic effects may also operate; for example rough fracture surfaces created by crack deflection mechanisms can result in additional crack bridging. This mechanical interlocking in the crack wake requires additional energy to overcome the frictional forces at the interlocking points.

(5) *Whisker pull-out*: whisker pull-out is also potentially a highly significant toughening mechanism. The mechanism requires relatively weak whisker/matrix interfacial bonding. As the whiskers pull out, the energy absorbed during crack progagation increases by the amount required to overcome the frictional forces at the whisker/matrix interface. The desired strength of the whisker/matrix interface is a function of whisker volume fraction, aspect ratio, modulus, and tensile strength and matrix modulus and strength. The whisker orientation is an important factor, highly oriented

whiskers normal to the crack plane being most effective. Control of the interfacial strength is critical for optimising toughening by whisker pull-out. Several intrinsic factors influence the nature of the whisker/matrix interface and limit the range over which interfacial strength can be controlled. If whiskers have smooth surfaces, then control of the chemical bond with the matrix by means of controlled reactions or coatings may provide interfacial strengths ranging from zero to the matrix strength. If whiskers have rough surfaces, then mechanical bonding will provide a substantial contribution to the interfacial strength. Mechanical bonding may also arise from residual stresses. When $\alpha_{whisker} < \alpha_{matrix}$, the whisker is under radial compression. This creates interfacial compressive stresses, which increase the effective shear resistance of the whisker/matrix interface. When $\alpha_{whisker} > \alpha_{matrix}$, the whisker is put in radial tension. This creates interfacial tensile stresses, which decrease the effective shear resistance of the whisker/matrix interface.

9.4.2   Fracture stress and fracture toughness

SiC whiskers have been successfully incorporated into numerous matrices. Matrix compositions have included Al_2O_3, Si_3N_4, mullite ($3Al_2O_3 \cdot 2SiO_2$, spinel ($Al_2O_3 \cdot MgO$), $MoSi_2$, ZrO_2 and glass ceramics.

Table 9.3   Summary of mechanical properties of SiC whisker/Al_2O_3 matrix composites

Whisker content (vol%)	Fracture strength (MPa)	Fracture toughness (MPa m$^{1/2}$)	Test temperature (°C)	References
5[a]	391	3.6	25	[29]
15[a]	652	4.6	25	[29]
0	150	4.3	25	[21, 19]
30[a]	680	8.7	25	[21, 19]
40[a]	850	6.2	25	[21]
40[a]	680	6.4	1000	[21]
40[a]	610	8.7	1200	[21]
0	–	4.5	25	[11]
10[a]	–	8.1	25	[11]
20[a]	–	7.6	25	[11]
30[a]	–	9.0	25	[11]
5[a]	475	4.0	25	[30]
10[a]	540	4.8	25	[30]
20[a]	670	6.1	25	[30]
30[a]	720	7.0	25	[30]
40[a]	640	7.9	25	[30]

[a] Silar-SC-9 Whiskers, Arco Chemical Co., Greer, SC.

Table 9.4 Summary of mechanical properties of SiC whisker/Si_3N_4 matrix composites

Whisker content (vol%)	Fracture strength (MPa)	Fracture toughness (MPa m$^{1/2}$)	Test temperature (°C)	References
0	780	4.7	25	[18]
30[a]	970	6.4	25	[18]
0	590	4.9	1000	[18]
30[a]	820	7.6	1000	[18]
0	490	6.2	1200	[18]
30[a]	590	7.7	1200	[18]
0	662	7.1	25	[12]
30[b]	450	10.5	25	[12]
0	900	6.0	25	[31]
10[c]	625	5.5	25	[31]
20[c]	575	5.0	25	[31]
0	375	4.0	25	[32]
10[a]	395	4.9	25	[32]
20[a]	550	7.0	25	[32]

[a]Silar-SC-9 Whiskers, Arco Chemical Co., Greer, SC.
[b]VLS Whiskers, Los Alamos National Laboratory, Los Alamos, NM.
[c]Tokamax Whiskers, Tokai Carbon, Tokyo, Japan.

Table 9.5 Summary of mechanical properties of SiC whisker/ceramic matrix composites

Whisker content (vol%)	Matrix material	Fracture strength (MPa)	Fracture toughness (MPa m$^{1/2}$)	Test temperature (°C)	References
0	Mullite	201	2.45	25	[33]
30[a]	Mullite	386	3.52	25	[33]
30[c]	Mullite	329	3.60	25	[33]
0	Mullite	–	2.2	25	[11]
20[a]	Mullite	440	4.6	25	[11]
0	ZrO_2	1150	6.0	25	[34]
20[c]	ZrO_2	600	10.5	25	[34]
30[c]	ZrO_2	600	11.0	25	[34]
0	$MoSi_2$	150	5.3	25	[35]
20[b]	$MoSi_2$	310	8.2	25	[35]
0	Alumino-borosilicate glass	103	1.0	25	[36]
35[a]	Alumino-borosilicate glass	327	5.1	25	[36]

[a]Silar-SC-9 Whiskers, Arc Chemical Co., Greer, SC.
[b]VLS Whiskers, Los Alamos National Laboratory, Los Alamos, NM.
[c]SCW-1 Whiskers, Tateho Chemical Ind., Kariya, Japan.

Tables 9.3–9.5 summarise measured fracture stress and fracture toughness values for these composites. It is apparent that fracture stress and
fracture toughness improvements are not consistently realised upon the
addition of whiskers in all cases. When improvements in fracture
toughness have been observed, the operative toughening mechanisms
have been related to whisker pull-out and bridging, crack deflection and
microcracking. Figure 9.3 illustrates these toughening mechanisms for a
SiC whisker/Al$_2$O$_3$ matrix composite. The variability in the expected
property improvements is related to several factors, including whisker/
matrix interfacial characteristics, whisker content and test temperature.
These factors are each discussed in the following sections. However, it

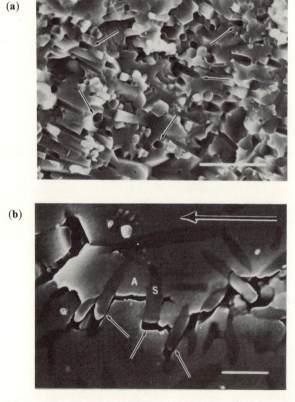

(a)

(b)

Figure 9.3 (a) Scanning electron microscope micrograph of fracture surface of a SiC
whisker (Silar SC-9 ARCO Metals Company)/Al$_2$O$_3$ matrix composite illustrating whisker
pull-out. Arrows indicate regions of whisker pull-out (bar = 5 μm) (Vaughn *et al.* [51]). (b)
Scanning electron microscope micrograph of crack-microstructure interaction of a SiC
whisker (Silar SC-9 ARCO Metals Company)/Al$_2$O$_3$ matrix composite illustrating crack
deflection and whisker bridging. Large arrow indicates direction of crack propagation and
small arrows indicate whisker bridging (bar = 2 μm) (Vaughn *et al.* [51]).

should be noted that processing defects (e.g. particulates and whisker bundles) can play an important role in limiting the fracture stress and fracture toughness. When processing defects are present, fracture toughness increases may still be realised, but fracture stress values may decrease.

Surprisingly little attention has been paid to the effect of whisker reinforcement on the scatter of strength. Rhodes *et al.* [37] report an increase in Weibull modulus from 4.6 to 13.4, with a corresponding increase in toughness from 3.7 to 6.7 MN m$^{-3/2}$, upon the addition of 15% SiC whiskers to an alumina. Further studies to confirm this trend would be desirable.

9.4.2.1 Effect of interfacial chemistry. Of particular importance in optimising the mechanical reliability of whisker reinforced composites is the effect of the whisker/matrix interfacial characteristics on the toughening and strengthing mechanisms. The degree of energy dissipation depends on the nature of the whisker/matrix interface, which is affected by several factors, including chemical bond formation, mechanical interlocking and thermal expansion mismatches. It is generally believed that a strong interfacial bond results in a composite exhibiting brittle behaviour, these composites usually have good fracture stresses but low fracture toughness. If the interfacial bond is weak, the composite will not fail in a fully catastrophic manner due to the activation of various energy dissipation processes. These composites tend to have high fracture toughness and low fracture stresses. In general, the interface should be strong enough to transfer the load from the matrix to the whiskers, but weak enough to fail prior to complete failure of the composite. Local damage then occurs without catastrophic failure. Thus, it is necessary to control the interfacial chemistry and bonding in order to optimise the overall mechanical reliability of the composite. This is a difficult task, considering densification processes occur at high temperatures.

The importance of the whisker/matrix interface has been illustrated by several investigators. Homeny and Vaughn [38] examined the effect of whisker surface chemistry on fracture stress and fracture toughness. Two SiC whiskers, Silar-SC-9 and Tateho-SCW-1-S, similar in all aspects except for surface chemistry, were used to reinforce polycrystalline Al_2O_3. The surfaces of the Tateho-SCW-1-S whiskers were found to consist essentially of SiC, while the Silar-SC-9 whiskers contained a substantial amount of excess oxygen and carbon in the form of SiO_xC_y. Table 9.6 illustrates the mechanical properties of the composites fabricated with these whiskers, along with a polycrystalline Al_2O_3. The composite fabricated with the Silar-SC-9 whiskers exhibited a significant increase in fracture toughness, while the composite fabricated with the

Table 9.6 Summary of mechanical properties of SiC whisker/Al_2O_3 matrix composites [38]

	Polycrystalline Al_2O_3	Composite (Silar-SC-9)	Composite (Tateho-SCW-1-S)
Percent theoretical density	99.2	100.0	99.2
Grain size (μm)	< 2.5	< 2.5	< 2.5
Fracture stress (MPa)	456 ± 40	641 ± 34	606 ± 146
Fracture toughness (MPa m$^{1/2}$)	3.3 ± 0.2	8.7 ± 0.2	4.6 ± 0.2

Tateho-SCW-1-S whiskers exhibited only a modest increase. Both composites, however, exhibited increases in fracture stress. Examination of fracture surfaces and crack/microstructure interactions revealed that the composites fabricated with the Silar-SC-9 whiskers exhibited extensive whisker pull-out and bridging, the pull-out and bridging lengths being of the order of 2–4 whisker diameters. Consideration of the thermal expansion coefficients of SiC [$4.7 \times 10^{-6}\,°C^{-1}$) and Al_2O_3 ($8.9 \times 10^{-6}\,°C^{-1}$) reveals that hoop tensions and radial compression exists in the matrix, while radial compression exists in the whisker, a residual stress state that increases the mechanical bonding at the whisker/matrix interface. The resultant effective shear stress, τ was estimated by Becher et al. as 800 MPa [21]. Estimating the critical whisker length, l_c, with the expression

$$l_c = (\sigma_{wF}r)/\tau$$

where r = whisker radius (0.3 μm) and σ_{wF} = whisker tensile strength (7 GPa), yields an l_c of approximately 2.5 μm, in agreement with the observed pull-out lengths. Apparently the thin layer of SiO_xC_y prevented a strong chemical bond from occurring.

Tiegs et al. [39] also examined the surface chemistry of Silar-SC-9 and Tateho-SCW-1 whiskers and correlated it to the bonding at the whisker/matrix interface and the fracture stress and fracture toughness of the composites. The surfaces of the Tateho-SCW-1 whiskers were found to be rich in SiO_2, while the surfaces of the Silar-SC-9 whiskers contained SiO_xC_y species. Composites fabricated with the Silar-SC-9 whiskers exhibited high fracture toughness (8.3 MPa m$^{1/2}$) and composites containing the Tateho-SCW-1 whiskers showed little enhancement in fracture toughness (4.2 MPa m$^{1/2}$). Examination of fracture surfaces revealed that high fracture toughness was associated with rough surfaces exhibiting extensive whisker bridging and pull-out, while low fracture toughness was associated with smooth surfaces. It was concluded that the high SiO_2 content on the surfaces of the Tateho-SCW-1 whiskers led to chemical bonding to the matrix and that the SiO_xC_y present on the

Silar-SC-9 whiskers prevented bonding, which allowed the observed toughening mechanisms to operate.

Numerous investigators have examined the structures of the whisker/matrix interfaces. Sarin and Ruhle [4] examined interfaces in SiC whisker/Al_2O_3 matrix composites using transmission electron microscopy. They found no evidence of interfacial reactions, but a continuous thin amorphous film was observed. The film was found not to be a function of the matrix chemistry, but rather the oxide rich surface of the whiskers. Lio et al. [30] investigated the interface with high resolution transmission electron microscopy. No evidence of an interfacial reaction was found, only the expected grain boundary mismatch region with a thickness of the order of 10–20 Å. Other investigators [17, 21], using transmission electron microscopy, have verified the lack of interfacial reaction products in the same composite system.

9.4.2.2 Effect of whisker content. A limited number of investigations document clearly the effect of whisker content on mechanical behaviour. In general, fracture stress and fracture toughness have been shown to increase with increasing whisker contents. These properties tend to reach a maximum for whisker contents of 30–40 vol% above which fabrication difficulties become an important factor affecting microstructural development. Table 9.3–9.5 illustrate that in most cases, increasing whisker additions result in incremental increases in both fracture stress and fracture toughness. In particular, the work of Wei and Becher [11] and Lio et al. [30] (Table 9.3), Buljan et al. [18] (Table 9.4), and Gac and Petrovic [35] (Table 9.5) illustrates these trends. As shown, whisker contents as low as 10 vol% have been shown to be effective in enhancing the mechanical reliability. In cases where whisker additions result in decreases in fracture stress, the detrimental effect is usually attributed to processing defects. In cases where whisker additions fail to produce enhanced toughening, improper control of whisker/matrix interfaces is usually considered the controlling factor.

9.4.2.3 Effect of temperature. A number of studies of SiC_w/Si_3N_4 [18] (Table 9.4) and SiC_w/Al_2O_3 composites [21, 40] (Table 9.3) indicate that the fracture behaviour remains linear-elastic up to about 1200 °C with little variation in fracture toughness or short term fracture strength. An apparent increase in toughness occurs in SiC_w/Al_2O_3 composites at around 1300 °C [40]. This is associated with the onset of sub-critical crack growth (*R*-curve behaviour) but no significant plastic or creep deformation. At 1400 °C in air, failure occurs by creep deformation and associated stable crack growth while at the same time oxidation becomes significant [40].

For extended times at high temperatures, oxidation of the non-oxide components becomes an important factor. Experiments by Porter *et al.* [29] revealed that for SiC whisker/Al_2O_3 matrix composites, exposure to air at elevated temperatures resulted in the formation of mullite ($3Al_2O_3 \cdot 2SiO_2$). The kinetics of the whisker/matrix reaction to form mullite were parabolic, indicating that oxygen diffusion across the mullite reaction product was rate controlling. Kriven *et al.* [41] examined the oxidation behaviour of SiC_w/Al_2O_3 and SiC_w/mullite matrix composites. The oxidation behaviour in air at elevated temperatures was similar to that of SiC and Si_3N_4 polycrystalline ceramics, where a protective layer forms. They also noted that for the SiC_w/Al_2O_3 composite, long time exposure resulted in a significant loss in fracture stress. The microstructural changes that accompanied the fracture stress degradation included the formation of mullite and SiO_2-rich glass at the expense of the SiC whiskers.

In addition to the effect of oxidation for extended times at high temperature on the mechanical behaviour, creep behavior also becomes a critical factor. Porter *et al.* [29], Chokshi and Porter [42], Porter and Chokshi [43], Xia and Langdon [44], and Lipetzky *et al.* [45] have performed extensive measurements on the creep behaviour of SiC whisker/Al_2O_3 matrix composites. In all cases, the addition of SiC whiskers significantly improved the creep resistance of polycrystalline Al_2O_3. The stress exponent, n, was calculated from the following form of the general creep relationship:

$$d\varepsilon/dt \propto \sigma^n$$

where $d\varepsilon/dt$ = steady state creep rate and σ = applied stress. Values of n between 4 and 5 were reported for the SiC whisker/Al_2O_3 matrix composites, as compared to n values between 1.5 and 2.0 for similar polycrystalline Al_2O_3 specimens. The n values between 1.5 and 2.0 have been attributed to diffusional creep mechanisms. It is clear that SiC whisker additions to polycrystalline Al_2O_3 altered the creep mechanisms, but the exact nature of these mechanisms is not clear. Proposed mechanisms include intragranular dislocation activity and accumulation of glassy phases at interface junctions followed by cavitation. Porter *et al.* [46] and Nixon *et al.* [47] have performed measurements on the creep behaviour of SiC whisker/Si_3N_4 matrix composites. They found that the addition of SiC whiskers had no affect on the creep resistance of polycrystalline Si_3N_4. This behaviour was attributed to the presence of amorphous grain-boundary phases originating in the Al_2O_3–Y_2O_3 sintering aid additions. Apparently, the large amount of glassy grain-boundary phase was the principal factor controlling creep. Observed stress exponents, between 0.5 and 2.0, tended to support the operation of viscous flow and/or diffusional creep mechanisms.

9.5 Physical properties

Experimentally measured properties for two SiC whisker reinforced alumina composites are given in Table 9.7. Since the composites are not isotropic, anisotropy of most of the properties is to be expected. However, the degree of whisker orientation and the orientation of measurement is not given for these data. Differences in orientation might account for some of the inconsistencies found when comparing the two composites.

It is to be noted that in contrast to unreinforced alumina both are electrically conducting indicating that the fibres form a continuous skeleton. A practical consequence of this is that the composite can be spark machined.

9.6 Applications

Whisker reinforced ceramics have potential for applications that demand hardness, wear resistance, chemical inertness and high-temperature capability. Their improved toughness implies a higher reliability than is offered by most monolithic ceramics. In air, the use of SiC_w is limited to about 1200 °C where oxidation becomes significant. The situation is not necessarily improved in inert atmospheres since in most situations at these temperatures SiC loses its passive protection and decomposes. The study by Karpman and Clark [2] shows that the production of whisker based composites is economically feasible. However, the environmental issues associated with the toxicity of ceramic fibres with sub-micrometre diameters have yet to be addressed adequately [48].

A proven specific application of a SiC_w/Al_2O_3 composite is as a cutting tool material, usually in the form of inserts for turning operations. In common with most other ceramic tool materials, the composite is most successful in the machining of Ni-base alloys at high rates.

Table 9.7 Selected properties of SiC whisker reinforced alumina composites

Property	15 wt% SiC [37]	33 vol% SiC[a]
Density (g/cm^3)	3.8	3.74
Hardness (Vickers, GPa)	20.5	20.6
Young's modulus (GPa)	425	395
Poissons ratio	0.24	0.23
Resistivity (ohm-cm)	1875	700
Coefficient of thermal expansion (K^{-1})(25–500 °C)	6.8×10^{-6}	6.0×10^{-6}

[a]WG 300, Greenleaf Corporation; manufacturers data.

Cutting speeds of up to 450 m/min with reasonable lifetimes can be achieved which represents a considerable improvement over earlier ceramics (see Figure 9.4).

Brandt *et al.* [49] have made a detailed comparative study of the machining of a low alloy steel and a Ni-alloy (Incoloy 718) with sialon and SiC_w/Al_2O_3 composite tool materials. This work provides some insight into the complex interplay of ceramic tool wear mechanisms and the machining parameters. The high toughness of both the tool materials means that the tool life or the attainable cutting speeds are not normally limited by fracture of the tool as is the case for non-toughened ceramics. For sialon the limiting criterion for tool life is instead flank wear (see chapter 4) resulting from reaction and interdiffusion between tool and workpiece. In the composite the flank wear is much lower apparently because of the formation of a Mg-rich oxide protective layer on the flank surface which inhibits interaction. The tool life at high speeds is thereby increased (Figure 9.4) and, instead of flank wear, is limited by depth-of-cut notch wear which occurs by a process of microfracture caused by the edge of the workpiece chip. The superior behaviour of the composite is not observed when machining steels. Significant flank wear occurs and in this study was associated with the dissolution of the SiC whiskers, presumably into the workpiece, as well as the penetration of the steel constituents into the ceramic.

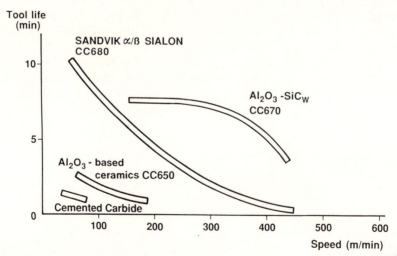

Figure 9.4 Comparison of the machining performance of various tool materials in cutting the Ni-alloy, Inconel 718 (by courtesy of G. Brandt, Sandvik Coromant AB, Stockholm). Feed 0.2 mm/rev; depth of cut 2 mm.

The source of Mg in the protective layer found when machining the Ni-alloy is thought to be the Mg used as a deoxidant in the alloy. It is also to be noted that the maximum temperature reached locally in the tool is about 200 °C higher when machining the Ni-alloy than when machining steels, and approaches 1200 °C at the higher speeds. The higher temperature implies that the workpiece will be softer and may also cause reaction layers to act as *in situ* lubricants.

9.7 Critical needs

There are a number of important areas where further research would greatly increase the probability of commercial use of whisker reinforced ceramics. These areas have been thoroughly documented by Lewis [27], Karpman and Clark [2], and Cornie *et al.* [50] and are summarised as follows:

(1) *Optimisation of whisker/matrix packing*: The whisker/matrix packing is a function of processing and material properties, for example whisker size and aspect ratio, matrix particle size and shape, relative whisker/matrix particle size, distribution of whisker sizes and matrix particle sizes, etcetera. More experimental and theoretical investigations are needed to tailor the whisker and matrix characteristics in order to achieve maximum preform density and appropriate homogeneous microstructures. This approach would enhance densification processes and would result in a maximisation of properties.

(2) *Interface control*: it is necessary to control the degree of matrix/whisker bonding and to prevent or minimise detrimental reactions or diffusion, which may degrade the whiskers. As indicated earlier, there exits a potential for controlling the nature and effectiveness of various toughening mechanisms through control of the interfacial bond and mechanical shear strength of the interface. Uncontrolled reaction or diffusion of matrix species into whiskers, which have small diameters, high surface areas and short diffusion distances, generally leads to degradation of the whiskers and severely restricts mechanical reliability. The ability to prevent or control whisker/matrix reactions and control interfacial bonding, possibly by means of coating techniques, is necessary because it permits a greater choice in the selection of whisker-/matrix combinations.

(3) *Whisker orientation*: there is a need for techniques to produce preforms with aligned whiskers. Possible processing routes include extrusion and spinning for orthotropic composites and infiltration

techniques for isotropic composites. These techniques may require some method of locking the whiskers together, such as an organic binder or a matrix precursor.

(4) *Net shape processing*: techniques for forming composites close to net shape in a cost effective manner are needed. New processing technologies must be developed to account for characteristics of components, interaction of components and features of final microstructures. This would include investigations for improved methods of densification, with a theoretical understanding of the mechanisms in a whisker/powder system.

(5) *Whisker handling and processing*: for whiskers produced from rice hulls, there is poor control over whisker size, aspect ratios, etcetera. Particulates and agglomerates are also common. Whiskers produced from the vapor-liquid-solid process are more amenable to processing and handling in a controlled manner. However, these whiskers suffer from significant contamination from the metal catalyst used in the whisker processing, which is not easily removed by either mechanical, chemical or thermal means.

References

1. Buljan, S. T., Pasto, A. E. and Kim, H. J. (1989) Ceramic whisker and particulate composites: properties, realibility, and applications, *Bull. Am. Ceram. Soc.* 68, 387–394.
2. Karpman, M. and Clark, J. (1987) Economics of whisker-reinforced ceramics, *Composites* 18, 121–124.
3. Milewski, J. V., Gac, F. D., Petrovic, J. J. and Skaggs, S. R, (1985) Growth of beta-silicon carbide whiskers by the VLS process, *J. Mater. Sci.* 20, 1160–1166.
4. Sarin, V. K. and Ruhle, M. (1987) Microstructural studies of ceramic matrix composites, *Composites* 18, 129–134.
5. Nutt, S. R. (1984) Defects in silicon carbide whiskers, *J. Am. Ceram. Soc.* 67, 428–431.
6. Nutt, S. R. (1988) Microstructure and growth model for rice hull derived SiC whiskers, *J. Am. Ceram. Soc.* 71, 149–156.
7. Homeny, J. and Vaughn, W. L. (1987) Whisker-reinforced ceramic matrix composites, *Mater. Res. Soc. Bull.* 10–11, 66–71.
8. Karasek, K. R., Bradley, S. A., Donner, J. T., Yeh, H. C., Schienle, J. L. and Fang, H. T. (1989) Characterization of silicon carbide whiskers, *J. Am. Ceram. Soc.* 72, 1907–1913.
9. Petrovic, J. J., Milewski, J. V., Rohr, D. L. and Gac, F. D. (1985) Tensile mechanical properties of SiC whiskers, *J. Mater. Sci.* 20, 1167–1177.
10. Schioler, L. J. and Stiglich, Jr., J. J. (1986) Ceramic matrix composites: a literature review, *Bull. Am. Ceram. Soc.* 65, 289–292.
11. Wei, G. C. and Becher, P. F. (1985) Development of SiC whisker reinforced ceramics, *Bull. Am. Ceram. Soc.* 64, 298–304.
12. Shalek, P. D., Petrovic, J. J., Hurley, G. F. and Gac, F. D. (1986) Hot pressed SiC whisker/Si_3N_4 matrix composites, *Bull. Am. Ceram. Soc.* 65, 351–356.
13. Phillips, D. C. (1983) Fiber reinforced ceramics, in: *Fabrication of Composites*, eds. Kelly, A. and Mileiko, S. T., North-Holland, Amsterdam, pp. 373–428.

14. Lee, K. W. and Sheargold, S. W. (1987) Particulate matters in silicon carbide whiskers, *Ceram. Eng. Sci. Proc.* 8, 702–711.
15. Lundberg, R., Nyberg, B., Williander, K., Persson, M. and Carlsson, R. (1987) Processing of whisker-reinforced ceramics, *Composites* 18, 125–127.
16. Hoffmann, M. J., Nagel, A., Greil, P. and Petzow, G., Slip casting of SiC whisker reinforced Si_3N_4, *J. Am. Ceram. Soc.* 72, 765–769.
17. Homeny, J., Vaughn, W. L. and Ferber, M. K. (1987) Processing and mechanical properties of SiC whisker Al_2O_3 matrix composites, *Bull. Am. Ceram. Soc.* 66, 333–338.
18. Buljan, S. T., Baldoni, J. G. and Huckabee, M. L. (1987) Si_3N_4–SiC composites, *Bull. Am. Ceram. Soc.* 66, 347–352.
19. Tiegs, T. N. and Becher, P. F. (1987) Sintered Al_2O_3–SiC whisker composites, *Bull. Am. Ceram. Soc.* 66, 339–342.
20. Becher, P. F. and Wei, G. C. (1984) Toughening behavior in SiC whisker reinforced alumina, *J. Am. Ceram. Soc.* 67, C267–C269.
21. Becher, P. F., Tiegs, T. N., Ogle, J. C. and Warwick, W. H. (1986) Toughening of ceramics by whisker reinforcement, in: *Fracture Mechanics of Ceramics, Vol. 7: Composites, Impact, Statistics, and High-Temperature Phenomena*, eds. Bradt, R. C., Hasselman, D. P. H., Evans, A. G. and Lange F. F., Plenum Press, New York, pp. 61–73.
22. Pezzotti, G., Tanaka, I., Okamoto, T., Koizumi, M. and Miyamoto, Y. (1989) Processing and mechanical properties of dense Si_3N_4–SiC whisker composites without sintering aids, *J. Am. Ceram. Soc.* 72, 1461–1464.
23. Bjork, L. and Hermansson, A. G. (1989) Hot isostatically pressed alumina-silicon carbide whisker composites, *J. Am. Ceram. Soc.* 72, 1436–1438.
24. Rice, R. W. (1981) Mechanisms of toughening in ceramic matrix composites, *Ceram. Eng. Sci. Proc.* 2, 661–701.
25. Rice, R. W. (1985) Ceramic matrix composite toughening mechanisms: an update, *Ceram. Eng. Sci. Proc.* 6, 589–607.
26. Shetty, D. K. (1982) Ceramic matrix composites, in: *Current Awareness Bulletin*, Metals and Ceramics Information Center, Battelle Columbus Labs, Vol. 118, no. 12.
27. Lewis, D. (1987) Whisker reinforced ceramics, in: *Processing of Advanced Ceramics*, eds. Moya, J. S. and Aza, S. D. Sociedad Espanola de Ceramica Y Vidrio, Madrid, Spain, pp. 49–72.
28. Warren, R. and Sarin, V. K. (1989) Fracture of whisker reinforced ceramics, in: *Composite Materials Series 6*, ed. Friedrich, K., Elsevier, Oxford, pp. 571–614.
29. Porter, J. R., Lange, F. F. and Chokshi, A. H. (1987) Processing and creep performance of SiC whisker reinforced Al_2O_3, *Bull. Am. Ceram. Soc.* 66, 343–347.
30. Lio, S., Watanabe, M., Matsubara, M. and Matsuo, Y. (1989) Mechanical properties of alumina/silicon carbide whisker composites, *J. Am. Ceram. Soc.* 72, 1880–1884.
31. Lundberg, R., Kahlman, L., Pompe, R., Carlsson, R. and Warren, R. (1987) SiC whisker reinforced Si_3N_4 composites, *Bull. Am. Ceram. Soc.* 66, 330–333.
32. Singh, J. P., Goretta, K. C., Kupperman, D. S. and Routbort, J. L. (1988) Fracture toughness and strength of SiC whisker reinforced Si_3N_4 composites, *Adv. Ceram. Mater.* 3, 357–360.
33. Samanta, S. C. and Musikant, S. (1985) SiC whiskers-reinforced ceramic matrix composites, *Ceram. Eng. Sci. Proc.* 6, 663–672.
34. Claussen, N., Weisskopf, K. L. and Ruhle, M. (1986) Mechanical properties of SiC whisker reinforced TZP, in: *Fracture Mechanics of Ceramics, Vol. 7: Composites, Impact, Statistics, and High-Temperature Phenomena*, eds. Bradt, R. C., Hasselman, D. P. H., Evans, A. G. and Lange, F. F., Plenum Press, New York, pp. 75–86.
35. Gac, F. D. and Petrovic, J. J. (1985) Feasibility of a composite of SiC whiskers in an $MoSi_2$ matrix, *J. Am. Ceram. Soc.* 68, C200–C201.
36. Gadkaree, K. P. and Chyung, K. (1986) Silicon carbide whisker reinforced glass-ceramic composites, *Bull. Am. Ceram. Soc.* 65, 370–376.
37. Rhodes, J. F., Dziedzic, C. J., Beatty, R. L and Cook, J. L. (1984) Ceramic reinforced ceramics, paper 43 in *PM Aerospace Materials, Proceedings of Metal Powder Report Conf. 1984*, MPR Publishing Services Ltd., Shrewsbury, UK.

38. Homeny, J. and Vaughn, W. L. (1990) Silicon carbide whisker/ alumina matrix composites: effect of whisker surface treatment on fracture toughness, *J. Am. Ceram. Soc.* 73, 394–402.
39. Tiegs, T. N., Becher, P. F. and Harris, L. A. (1987) Interfaces in alumina-SiC whisker composites, in: *Ceramic Microstructures 86*, eds. Pask, J. A. and Evans, A. G. Plenum Press, New York, pp. 911–917.
40. Han, L. X., Hansson, T., Suresh, S. and Warren, R. (1989) High temperature fracture of SiC whisker reinforced alumina, in: *Proceedings of the 5th Scandinavian Symposium on Materials Science*, eds. Hansson, I. and Lilholt, H., Danish Society for Materials Testing and Research, Copenhagen, pp. 287–294.
41. Kriven, W. M., Van Tendeloo, G., Tiegs, T. N. and Becher, P. F. (1987) Effect of high temperature oxidation on the microstructure and mechanical properties of whisker reinforced ceramics, in: *Ceramic Microstructures 86*, eds. Pask, J. A. and Evans, A. G., Plenum Press, New York, pp. 939–948.
42. Chokshi, A. H. and Porter, J. R. (1985) Creep deformation of an alumina matrix composite reinforced with silicon carbide whiskers, *J. Am. Ceram. Soc.* 68, C144–C145.
43. Porter, J. R. and Chokshi, A. H. (1987) Creep performance of silicon carbide whisker-reinforced alumina, in: *Ceramic Microstructures 86*, eds. Pask, J. A. and Evans, A. G., Plenum Press, New York, pp. 919–928.
44. Xia, K. and Langdon, T. G. (1988) The mechanical properties at high temperatures of SiC whisker-reinforced alumina, in: *High Temperature/High Performance Composites*, eds. Lemkey, F. D., Fishman, S. G., Evans, A. G. and Strife, J. R. Materials Research Society, Pittsburgh, pp. 265–270.
45. Lipetzky, P., Nutt, S. R. and Becher, P. F. (1988) Creep behavior of an Al_2O_3–SiC composite, in: *High Temperature/High Performance Composites*, eds. Lemkey, F. D., Fishman, S. G., Evans, A. G. and Strife, J. R., Materials Research Society, Pittsburgh, pp. 271–277.
46. Porter, J. R., Lange, F. F. and Chokshi, A. H. (1987) Processing and creep performance of silicon carbide whisker-reinforced silicon nitride, in: *Advanced Structural Ceramics*, eds. Becher, P. F., Swain, M. V. and Somiya, S., Materials Research Society, Pittsburgh, pp. 289–294.
47. Nixon, R. D., Chevacharoenkul, S., Huckabee, M. L., Buljan, S. T. and Davis, R. F. (1987) Deformation behavior of SiC whisker reinforced Si_3N_4, in: *Advanced Structural Ceramics*, eds. Becher, P. F., Swain, M. V. and Somiya, S., Materials Research Society, Pittsburgh, pp. 295–302.
48. Bogoroch, R. and Luck, S. R. (1988) Workplace handling requirements and procedures for ACMC SiC whiskers based on subchronic inhalation study in rats, in: *Whisker and Fiber Toughened Ceramics, Proceedings of International Conference*, Oak Ridge, TN, eds. Bradley, R. A, Clark, D. E., Larsen, D. C. and Stiegler, J. O., ASM International, pp. 81–89.
49. Brandt, G., Gerendas, A. and Mikus, M. (1991) Wear mechanisms of ceramic cutting tools when machining ferrous and non-ferrous alloys, to be published (authors' address: AB Sandvik Coromant, Stockholm, Sweden).
50. Cornie, J. A., Chiang, Y. M., Uhlmann, D. R., Mortensen, A. and Collins, J. M. (1986) Processing of metal and ceramic matrix composites, *Bull. Am. Ceram. Soc.* 65, 293–304.
51. Vaughn, W. L., Homeny, J. and Ferber, M. K. (1987) Mechanical properties of silicon carbide whiskers/alumina oxide matrix composites, *Ceram. Eng. Sci. Proc.* 8, 848–859.

List of symbols

Note: f, m, p and c subscripts generally denote fibre, matrix, particle and composite, respectively, unless otherwise stated.

a	notch or pre-crack length in a fracture toughness test
A	area
A_{ii}	extensional stiffness
B	specimen thickness in bend test
B_{ii}	coupling stiffness
c	size of defect or flaw, etc.
C	compliance
C_p	specific heat
D	gas flow rate
D_e	effective diffusion coefficient
D_F	Fick diffusion coefficient
D_K	Knudsen diffusion coefficient
D_0	pre-exponential constant for diffusion coefficient
D_{ii}	bending stiffness
E	Young's modulus
G	(i) contiguity of second phase particles or (ii) shear modulus
G_I	strain energy release rate for mode I crack
G_{Ic}	critical strain energy release rate
h	height or width of crack process zone
H	hardness
k_0	pre-exponential constant for k_s
k_s	surface reaction constant
K, K_t	thermal conductivity
K_c	critical stress intensity factor of unspecified mode
K_Q	apparent or candidate fracture toughness
K_r	apparent fracture toughness (in rupture of CVI composites)
K_I	mode I stress intensity factor
K_{Ic}	mode I critical stress intensity factor, fracture toughness
l	length
l_c	critical length (of fibre)
l_t	transfer length
L	length
m	Weibull modulus or exponent
M	moment
M_p	mass of ceramic deposited in pores during CVI
M_s	mass of ceramic deposited on an external surface during CVI
n	(i) number or (ii) exponent
N	(i) number or (ii) force
P	(i) pressure or (ii) load
P_f	probability of failure below a given stress
P_s	probability of survival of a given stress
q	exponent
Q	activation energy
Q_{ii}	stiffness coefficient
r	radius
r_p	pore radius
R	gas constant
S	local internal stress
S_{ii}	compliance coefficient
S_{mp}	average fractional area of contact or particle with matrix

S_{pp}	average fractional area of contact of particle with neighbouring particles
t	time
T	temperature or transformation matrix
u	crack-opening displacement
V	volume
v_m, v_p, etc.	volume fraction
V_m, V_p, etc.	volume fraction
V_0	normalising volume constant in Weibull equation
V_p	volume fraction (i) particles or (ii) porosity
V'_{po}	volume fraction porosity preform prior to infiltration
W	(i) wear rate or (ii) specimen width in bend test
x	distance
X	dimensionless constant $= k_s r_p / D_e$
Y	dimensionless constant
z	distance
α	(i) coefficient of thermal expansion or (ii) molar ratio of gaseous precursors in CVI
γ	specific fracture surface energy
Γ	specific fracture surface energy
ΔT	change in temperature
ε	strain
$\dot{\varepsilon}$	strain rate
ε_F	fracture strain
η	viscosity
κ	thermal diffusivity
ν	Poisson's ratio
ρ	density
ρ_l	density of a liquid
ρ_s	density of a solid
σ	stress
σ	mean stress
σ^*	stress in one phase at fracture strain of another phase
σ'_f	maximum stress along the length of a fibre
σ_F	fracture stress
σ_0	normalising constant in Weibull equation
τ	interfacial shear or friction stress
υ	sedimentation velocity
ϕ	thermal shock resistance (figure of merit)
ω	electrical resistance

Index

£60